FUNDAMENTAL BUILDING MATERIALS

FUNDAMENTAL BUILDING MATERIALS

Fourth Edition

Ken Ward-Harvey, ASTC, LFRAIA

Universal-Publishers
Boca Raton

Fundamental Building Materials
Fourth Edition

Universal-Publishers
Boca Raton, Florida • USA
2009

ISBN-10: 1-59942-954-3/ISBN-13: 978-1-59942-954-0 *(paper)*
ISBN-10: 1-59942-951-9/ISBN-13: 978-1-59942-951-9 *(ebook)*

www.universal-publishers.com

Cover design by Shereen Siddiqui

First published 1984
Sakoga Pty Ltd.

Second Edition, 1989, Third Edition, 1997
The Royal Australian Institute of Architects

Fourth Edition reset 2008
WHO Presentations Services NSW

Library of Congress Cataloging-in-Publication Data

Ward-Harvey, K.
  Fundamental building materials / Ken Ward-Harvey. -- 4th ed.
     p. cm.
  Includes bibliographical references and index.
  ISBN-13: 978-1-59942-954-0 (pbk. : alk. paper)
  ISBN-10: 1-59942-954-3 (pbk. : alk. paper)
  1. Building materials. I. Title.
  TA403.W335 2009
  691--dc22
                              2009006519

ACKNOWLEDGEMENTS

To my wife Joan for her loyalty and support.

To Marjorie McNeece, Shirley Mathes and Mabs Watkins whose patience and assistance with editing and typing made this book possible.

To Ken Wyatt, Gordon Renwick, Vernon Ireland and Jack Greenland of NSWIT who helped add detail to some sketchy outlines.

For assistance with photography and illustrations I thank Paul Finigan, Adrian Boddy, Peter Elliott, Cement & Concrete Association of Aust, Public Works Dept Victoria, The Manly Daily, Building Materials and Equipment Australia and New Zealand, NSW Forestry Commission.

To students and other readers whose favourable comments helped me persevere.

To the RAIA Architects Advisory Service for support and assistance with publication.

To manufacturers and publishers who willingly permitted reproduction of their material.

Angus & Robertson
ACI Fibreglass
Australian Surfacing Contractors
Australian Lead Development Association (ALDA)
Austral Bronze Crime Copper
BASF Australia Ltd
Brick Development Research Institute (BDRI)
Broken Hill Associated Smelters (BHAS)
Cement & Concrete Association of Australia (CACA)
Comalco Aluminium
Dow Corning (Aust) Pty Ltd
Duncans - Sawmillers & Timber Agents
Experimental Building Station, North Ryde (EBS)
Hardboards Aust. Pty Ltd
Hardie-Iplex Plastics
John Lysaght (Australia) Ltd
Monier Ltd
National Capital Development Commission
NSW Forestry Commission
Pilkington-ACI Pty Ltd
Plywood Association of Australia
St Regis-ACI Pty Ltd
Sulphide Corporation Pty Ltd
Standards Association of Australia (SAA)
Timber Development Association (TDA)
Vessey Chemicals
Zinc: Today

1997 Edition

Since the first edition in 1984 many changes have taken place to the above, and the following additional acknowledgements for help received should be recorded :
Bonnie Ward-Harvey
H.P. James FNZIA
Robert Macindoe & Mark Fenwick of Suters Architects, Newcastle
Jim Sutherland - Mr Ply&Wood
Bill Barbas - BHP Steel Direct
Boral Besser Masonry & Timber
Clay Brick & Paver Institute
CSIRO Division of Building Construction & Engineering
CSR Building Materials
Forwood Products
NSW State Forests
Plywood Association of Australia
Standards Australia

2009 Edition
Publications
Architectural Product News by Reed Business Information
Bluescope Lysaghts Referee
Corrosion Management by Industrial Galvinisers
Kingspan Insulated Panels
James Hardie - The Smarter Construction Book
              - The Smarter Green Book
Viridian New World Glass. Architectural Specifiers guide.
The Architects Handbook 2007/08. RAIA
Andrew Arnott from Port Stephens Telecentre

FOREWORD

People involved in the building industry are confronted with a great array of materials made from natural resources such as timber, and manufactured articles such as bricks and metals. Many of these have been used effectively for centuries, but increasingly newly developed materials and systems for construction come on the market and need to be evaluated before they can be effectively incorporated and stand the long term uses required of buildings. Some proven materials can become unpopular due to labour costs, or banned due to public health risks, as with lead based paints and asbestos fibres in the latter half of the 20th Century.

Traditionally much building materials knowledge was acquired slowly within the apprenticeship or pupilage systems, but modern commercial pressures and rapid changes in the 20th century have shown these systems to be too slow. This book has been prepared by an architect with fifty years experience of practice and teaching, in an attempt to provide a wide ranging introduction to this very broad subject. Emphasis is placed on visual identification of materials through photographs of typical products. Understanding can only be developed once visual identification is established, and this is basic to all communication within the industry. Personal observation of material uses, performance and case studies is essential.

The Building Code of Australia (BCA) was introduced in 1988 to supersede the various State Building Regulations, and simplify interpretation of the rules for manufacturers and people who often work across State boundaries.
Amendments are still being added. Due to the wide range of Australian climatic conditions from cyclonic winds, heat and humidity; to dry desert; alpine snowfields; and the favoured pleasant coastal littoral, some special clauses had to be included to cover these extremely varied, conditions. Generally speaking Australia now works to a set of National methods of construction. We also have available a Standard Specification called NATSPEC.

This book should also become a desk reference for professionals; as further study is assisted by the summaries and technical references listed for each chapter. While the majority of these are of Australian origin, British, New Zealand and other publications are mentioned. The rapid exchange of goods and information now possible has led to many building materials being transported across the world, and foreign research and Standards being used in their assessment.

Serious concerns have been formally expressed by Australian national bodies involved with building performance and maintenance regarding widespread evidence of failures and breakdowns due to lack of understanding of material characteristics. Such understanding should be a basic requirement for individual tradesmen or sub-contractors regarding their materials generally, as well as for the architect and builder.

The thrust of this book is to create a critical approach toward learning about materials as the basis for personal understanding and development. This method, aided by research reports and other published information regularly coming forward, should be universally applicable, and serve well at the professional, trade and student level.

Ken Ward-Harvey.

September 2008

CONTENTS

PREFACE TO 2009 EDITION

The history of Architecture and Building goes back thousands of years, and the major changes termed "styles" have been closely associated with the materials available and the technologies learnt by tradesmen to construct those outstanding buildings which have come down to us as great examples of past Architecture. The materials were mainly timber or derived from plants ; stones, and ceramics such as bricks and other forms of baked clay products. Changes were usually slow and easily passed on by the tradesmen, who were largely limited to Carpenters, Stonemasons, and Bricklayers.

With the industrial revolution many more materials were produced which could be beneficially incorporated into buildings. By 1851 the revolutionary 'Crystal Palace' in London was built from cast iron, glass and timber in very fast time. That and other buildings to follow, changed the concept of on site hand work for building, to using factory made products such as steel, and its companion reinforced concrete, so a new style of Architecture began to emerge. The range of materials to be understood by designers and master craftsmen was greatly expanded during the late 19th & 20th centuries, as many new materials and technologies were introduced and tested. So called "Modern" architecture of the 20th century bore little resemblance to building forms of previous centuries. The range of materials is still expanding, consuming increasing quantities of energy, and further polluting our atmosphere.

In the ten years since the text of this book was last revised there have been great changes in the building industry's attitudes, knowledge and communications, which led to a new approach in specifying the roles of many materials. This has come about due to the increasing awareness of "climate change"; the impact of fossil fuel use on our climate and weather; and improved methods of harnessing wind and solar power; the concept of "Sustainability" of building construction; and the tremendous impact of electronic communications, especially the Internet and its offshoot E-mail. These have enabled manufacturers to place information regarding their products in this publicly accessible forum, which can be periodically updated. Consequently many references included in previous editions are now not included as they can become quickly outmoded.

Our planet is at a crisis point which must be acknowledged, especially by those involved in construction industries, which consume large quantities of materials and power in mining, harvesting, manufacture and construction, and continue to consume power throughout the life of the buildings or other works created.

Designers and Craftsmen create the future, and their decisions determine a building's impact on the environment over much of its entire life. The right decisions cost less during the design & construction phase than at any other point in a building's life cycle. So the choices made by designers & specification writers regarding chosen materials is critical. Energy consumed in the manufacturing processes, and performance of the -proposed built elements as shelter from heat; cold; noise, sunlight, and power consumption must all be considered. Such scientific qualities need to be understood and used by designers. Selection of materials in the 21st century has become more scientific than in the past, which was largely a choice in pursuing local 'styles' with a very limited range of materials to use. This is no longer valid for those claiming professional status and expertise in the building industry.

Major material manufacturers are now aware of the potential benefits of the Internet and update technical information on their products. It is not necessary now for a book such as this to record many static lists and formal details, which may shortly become out of date. Students should seek out current printed material and read it in conjunction with the introductory material herewith. Some materials are now available as factory assembled to simplify on site applications; e.g. Insulation on roof and wall panels. By using these revised References together with Trade Literature the reader should gain a much clearer picture of how the elements of buildings perform, and expectations for to-day and tomorrow's requirements. Remember Standard Codes as well as the Building Code of Australia carry legal weight if disputes regarding buildings reach Court.

The Architects Handbook and the BEDP Environment Design Guide published bi-annually by the Royal Australian Institute of Architects (RAIA), contain articles and information supplementary to much contained herein, including many useful addresses. Hints regarding Specifying for Sustainable Building; Climate & Weather Information; Cost Guide; Roofing & Guttering; Cracking Brickwork, and many other common problems are included. Some major manufacturers have produced excellent booklets and pamphlets available on application.

The basic text of this book has served well for over twenty years and is left much as previously, except for minor alterations. A guide to some additional and/or revised information and references now available has been listed at the end of some chapters.   Ken Ward-Harvey. September 2008

INTRODUCTION

*The need; traditional and current knowledge; traditional skills; contemporary problems;
present approach; terminology and references.*

Anyone involved in a responsible role in building needs a very broad understanding of a wide variety of materials, their potential and deficiencies in use. The aim of this book is to provide this fundamental understanding as a starting point.

Many formal courses in architecture and building teach building construction as a major subject and refer to the products used with little background information regarding these, their manufacture, raw materials and peculiarities.

When local buildings used local products, which were basically few, easily identified and understood by local tradesmen, this was a tolerable situation. Present-day conditions in industrialised communities are very different.

Materials of exceptional qualities have often been transported long distances for special projects, but it was not until the 19th and 20th centuries that modern transport made this commonplace for modest houses as well as major monuments.

As a result of this, the variety of materials available in cities and many industrialised regions has increased dramatically. Whereas stone, brick, timber, mortar, plaster, terra cotta and slate were the primary materials two hundred years ago, the range commonly available today, even under those general classifications, is far wider than ever before. With the rise of the steel, cement, aluminium, glass and chemical industries, comparative newcomers as major building component manufacturers, the traditionally restricted range has exploded.

To help readers extend their knowledge and to relate the material characteristics to further studies of its uses, references are given to many documents from recognised Australian authorities and some overseas publications. Trade publications referred to often have local counterparts in other countries.

Local climatic conditions vary considerably and materials perform differently according to climatic exposure. Coastal conditions are more corrosive on many materials than dry inland atmospheres. Ultra-violet intensity varies throughout the world and Australian conditions are particularly harsh in this regard.

Students and practising professionals in the building industry have also to be in touch with buildings and site works, factories, fabrication workshops, building information centres, manufacturers' literature, research and testing organisations, to build up their knowledge and understanding.

They can help themselves and the construction industry as a whole by requesting that test reports be supplied on any new materials or assembly, preferably from a locally recognised authority; and by knowing which properties are likely to be the critical ones in any particular situation in which they are choosing building materials, systems or finishes. This book should assist greatly in this regard.

Careful observation of completed buildings for successful or unsuccessful materials, details and finishes and further extensive reading is an essential part of this learning process, and must continue throughout a professional career.

TRADITIONAL AND CURRENT KNOWLEDGE

Materials science has now advanced far beyond the levels of knowledge which existed at the great periods of historical building, but the performance and characteristics of materials which people in the building industry need to know about are still much the same and can be identified as primarily those pertaining to:

(a) resistance to structural stress
(b) process of manufacture
(c) effects of water, freezing and thawing, and atmosphere
(d) heat and temperature effects on material/product
(e) effects of ultra-violet radiation
(f) electrolytic or other special characteristics
(g) acoustic properties.

Obviously, the foregoing list relates mainly to physical characteristics, but it includes some small relationship to the chemical and electrical sciences, particularly when long-term durability of materials subject to contaminated environments or freezing are a consideration. Even biological sciences become involved with natural materials such as timber.

Atomic physics, molecular structure and such topics all form part of this total performance, but to the man in the building industry without a strong grounding in formal science, the simplest possible explanation of cause and effect is usually the best starting point. For problem situations, experts should be consulted as quickly as possible.

Traditional Skills

Most traditional building materials, such as brick, stone, timber, copper and lead, have served man well for many centuries. Craft skills, embodying much of the wisdom and knowledge gained from past experience working these materials, were handed down via the apprenticeship system and the trade guilds.

Thus tradesmen could be reasonably sure of the performance of their work in providing a satisfactory weather-resistant building. Designers and contractors knew they could rely on the tradesmen to produce quality work.

Master craftsmen, who developed from such training and experience, often became the building designers, even for such great works as the European and English cathedrals of medieval times. These master craftsmen were in great demand and some travelled widely to advise on then current building projects. Despite local variations of materials and skills, the breadth and depth of their craft knowledge was generally transferable and adaptable to new situations as they arose. The many great buildings designed by such people, which are still admired for their artistic and structural qualities, are evidence of the value of sound craft-type knowledge as a basis for architectural achievement, when only a few materials are used.

Contemporary Problems

Since the development of industrialised manufacturing processes for iron, steel, cement, aluminium and glass in the 19th and 20th centuries, the range of materials used in building structures and finishes has expanded enormously. The traditional accumulated knowledge, understanding and wisdom handed down through the apprenticeship system have been lacking, due largely to inability of the old personalised training methods being able to survive under commercial pressures.

Many modern buildings have developed serious defects, because certain combinations or characteristics of materials used have not been understood by the designers. Responsibility for the choice of major materials, jointing and assembly methods is now very much with the architect, builder or structural engineer. Obviously these professionals cannot also pursue numerous trades courses, but they need to have a strong understanding of the relevant physical properties and behaviour of many building materials. If their knowledge of materials and systems is weak or not up to date, this will inevitably result in defects on any large scale job which could have very serious consequences, both for the professionals and the building.

Present Approach

This book aims to provide a general descriptive background for commonly encountered building materials so that the student can develop a vocabulary regarding materials which will allow for the gradual development of deeper understanding.

To pursue a study of building materials it is convenient to deal in categories which have distinct relationships, usually derived from their raw material source.

INTRODUCTION

The categories we shall deal with in this book are as follows:
Water, its solid, liquid and vapour states
Foundation materials
Primitive materials
Timbers and timber products
Bricks and brickwork
Ceramics and/or pottery
Building stones and stone products
Limes, cements, mortars, plasters, renders
Concrete, cement and concrete products
Metals - ferrous and non-ferrous
Glass, glass fibres arid glass products
Chemically based products
Incompatibility of materials
Comparative tabulations
The sequence is not critical and chapters may be taken in any order.

It is impossible to deal comprehensively with the subject of building materials in an architecture, building or engineering course as it is now a specialised branch of science to which new and even radical developments are frequently being introduced with varying degrees of success. Obviously sincere efforts must be made to develop a satisfactory level of knowledge regarding materials because of the peculiar relationships which occur between differing materials in building situations and the lengthy life expectancy required.

Formalised technical references are nominated where relevant to assist in pursuing further the industrial information available.

Terminology & References

The terminology used is that common to the building industry in Australia, where only minor variations occur between States. Recent compilations of common terms have resulted in the SAA HB 501994 Glossary of Building Terms which contain useful illustrations.

The Standards Association of New Zealand has some very useful publications

1. Glossary of Building Terminology; which is an alphabetical listing plus some illustrations.

2. Standards Catalogue; which also contains some names of British and other Standards where relevant. The alphabetical index is probably the easiest to use of such Catalogues. Australia and New Zealand now generally co-operate on development of standards of which there are over 2500 AS/ NZS codes serving both countries. Key Australian and joint codes are now available on a CD.

The major abbreviations used refer to the following :-

AS   *Australian Standard produced by Standards Australia,*
         286 Sussex St, Sydney 2000
BS   *British Standard produced by the British Standards*
         Institution 389 Chiswick High Road London W4 4AL UK.
NZS   *New Zealand Standard produced by the SANZ;*
         Standards House 155 The Terrace Wellington 6020 NZ
CSIRO   *Commonwealth Scientific & Industrial Research*
         Organization Division of Building Construction and
         Engineering PO Box 56 Highett Victoria 3190 Australia
NSB   *Notes on the Science of Building published by the CSIRO*
BHP   *BHP Billiton 600 Bourke St, Melbourne 3000*
         E-mail STEELWOL.CustomerSupport@blzp.com.au
CB&PA   *Clay Brick & Paver Association NSW*
         PO Box 569 Wentworthville 2145
TDA   *Timber Development Association*

More detailed historical background regarding building materials is available in the "Macmillan Encyclopaedia of Architecture and Technological Change" edited by Pedro Guedes 1979, and from articles under appropriate headings in major Encyclopaedia.

Further technical background on materials can also be obtained from AJ Handbooks of `Building Structure' and `Building Enclosure', which should be in most technical libraries.

Standards and Codes

Many Countries and cities now have well developed building standards or codes. These have usually been argued over and developed by committees containing members of the relevant professions, trades and manufacturers involved with these products, and should be respected by all in the local building industry. The best way to learn about these in your particular sphere of interest is to find them on the internet, and purchase copies of codes if possible.

The following list of Australian & New Zealand standard codes indicates the sorts of topics covered. There may be similar ones for your local situation. Lists appearing at the end of chapters, may not be relevant outside of the Australian & New Zealand spheres.

General Standards

HB 90.3-2000 Guide to the construction industry
SAA HB 50:2004 Glossary of building terms

AS ISO 717.1:2004 Acoustics Airborne sound insulation
AS ISO 717.2:2004 Acoustics Impact sound insulation
AS/NZS 1170.0-1-2-3:2002 Structural Design. General principles
ASH 70.4:1993 SAA Loading Code, Earthquake loads
AS 1428.1:2001. Design for Access. New building work
AS 1428. 4:1992 Design for Access for the Vision Impaired
AS 470.1986 Health & safety at work
AS 1530.4;1997 Fire tests of building elements
AS 657:1992 Fixed Platforms, walkways, stairways, stairs etc
AS 1668.2-2002 Ventilation & Air Conditioning in buildings
AS/NZS 1680.0-1:1998 Interior lighting - Safe movement
AS 1905.1-2005 Fire resistant components
AS 2293.1-2005 Emergency escape lighting & exit signs
AS 2419.1-2005 Fire Hydrant installation & commissioning
AS/NZS 3000: 2000 Australian /New Zealand wiring rules
AS/NZS 3661.1 &2 Slip resistance of pedestrian surfaces
AS 3959:1999 Buildings in bushfire prone areas
AS 3999:1992 Bulk Thermal insulation of dwellings
AS 4005-2006 Wind loads for housing
AS 4226.1994 Guidelines for safe housing design
AS 4859.1:2002 Materials for thermal insulation
AS 5601-2004 Gas Installation
AS/NZS.ISO. 9001.2000.Quality Management Systems

Radiation Protection

British and Australian Standards also cover numerous topics of importance in various specialised scientific, industrial, and medical situations where X-Rays and/or radioactive materials are encountered.

Acoustics

Australian Codes also include some similarly general publications especially regarding acoustics, which are dealt with in considerable detail. The SAA Building Index should be referred to for the complete list.

WATER

Liquid, solid, vapour - Moisture Content - Chemical Combinations, Electrolytic action, In construction process, Damaging effects, Underground water.

Introduction

Water is an element of extremely important consideration in all building design and construction.

Protection from rainwater is one of the primary reasons for the construction of most buildings, and water in some of its physical forms and combinations with other chemicals also influences the life and efficiency of many building materials, components and details.

The Three States of Water

Chemically, water is a compound of hydrogen and oxygen ($H2O$), commonly encountered in various physical states. It can combine with various other chemicals to form dilute acids or alkalis.

The three states of water normally of concern in the design and maintenance of buildings are:
(1)   ice - solid frozen water, includes snow
(2)   water - liquid, usually derived from rainwater
(3)   water vapour - steam and moisture in air

The physical changes between these states are caused primarily by temperature changes frequently encountered in normal climatic variations.

Ice

As ice occupies greater space than does its equivalent quantity of water, expansion occurs on freezing. This can damage or force apart materials penetrated by the original liquid. Freezing conditions can also burst water pipes.

Liquid

Water in its liquid form will normally flow downwards under gravitational influences, but it can also move upwards under wind pressure, and horizontally or vertically by capillary attraction through porous or semi-porous materials.

When water (or any liquid) is contained so that it builds up in depth, the pressure at the base of the containment increases proportionally to the depth. In pipes, retaining walls, basements, etc. this water pressure can become a serious structural as well as a waterproofing problem. The water pressure developed is often referred to as the 'hydrostatic head' and is expressed in kilopascals (kPa).

Vapour

Water vapour can exist invisibly in air (humidity) and can penetrate almost any open part of a building, sometimes depositing itself as droplets on a surface cold enough to cause vapour fallout. This is the phenomenon frequently observed in winter on glass windows and sheet metal surfaces and known as condensation.

MOISTURE CONTENT

Many porous materials have the ability to absorb liquid or vapour with varying degrees of effect on their physical and structural characteristics, depending on their material of origin, manufacturing process, etc.

The water contained in a material is usually referred to as the 'moisture content' and is measured as a percentage of the dry weight of the material.

Timber

Organic materials such as timber are originally very moist when first cut and need to be dried out or 'seasoned' to attain their maximum strength and desirable characteristics for construction purposes. Variations from this desirable moisture content after seasoning (which varies with different climatic locations) leads to swelling for moisture added, or shrinkage for moisture deducted from the optimum. Some results of this are frequently seen in long periods of high humidity, when doors bind on their frames. Dry, windy weather will often cause joinery to shrink and open up cracks at joints.

Clay products

Clay and most of its products are hygroscopic; that is, they can absorb water unless glazed in the firing processes. In its natural state clay changes dramatically from hard to plastic according to the moisture content. Kiln-fired clay building products such as terra cotta, bricks and tiles do not usually change so dramatically but can absorb considerable quantities of water which add to the dead weight of the dry materials and reduce effectiveness for weatherproofing and some forms of insulation.

Cement products

Cement and concrete products require the introduction of water to create the chemical reaction which binds the ingredients together and this water has to dry out of the product, during which process a degree of shrinkage takes place. Once dry, however, moisture can again be absorbed and when this occurs some expansion takes place in the product. In situations where straight walls, etc. are built to close tolerances, this expansion can be sufficient to cause unsightly deformation and cracking of masonry or components affixed to the cement and concrete products.

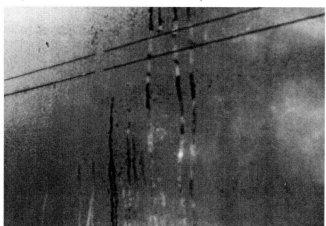

Condensation formed on a glass door due to internal humidity and exterior colder temperature
Condensation under timber flooring due to dampness from ground and inadequate ventilation. This causes decay of timbers.

CHEMICAL COMBINATIONS WITH WATER

Many minerals used in building have the ability to combine with water chemically to form compounds which can be either beneficial or detrimental to a building depending on whether the combination is deliberate or accidental.

Some of the obviously beneficial uses are plasticising and hydrating. limes, plasters, cement etc. Rusting of steel, erosion of masonry, etc. are familiar building defects brought about by chemical and/or physical combinations.

Water and atmospheric pollution

The atmosphere close to the sea coast absorbs quantities of sea salts, and locations in other areas similarly carry industrial pollutants which can be deposited on building surfaces. When dissolved by rain these produce dilute acidic or alkaline mixtures which attack the surfaces of materials or penetrate cracks or joints.

Corrosion has been defined for architecture, engineering and industry as the undesirable deterioration of constructional materials by electrochemical action resulting in loss of functional or aesthetic value. (IMP Coating Systems Data Sheet 2001-1, January 1969.)

(See also NSB No. 79 on Corrosion of Metals in Building, and Building Research Station Digest No. 98 - Durability of Metals in Natural Waters.)

Effects of this type of moisture penetration and corrosion are readily visible on many concrete lintels and arch bars in coastal locations where the steel has been rusted causing large brickwork cracks and spalling of concrete.

Similarly, galvanised steel sheeting is more rapidly deteriorated near the coast than inland because of the high level of atmospheric salts which absorb water vapour and maintain a constant corrosive attack on the metals.

Soluble salts often penetrate masonry and other materials in water then form crystals as the water evaporates, causing delamination, cracking and splitting. (See chapter on Incompatibility of Materials.)

Electrolytic or Galvanic action

Water is often involved in electrolytic or galvanic action when two materials (usually metals) which are widely separated on the `galvanic table' come into close contact.

The natural differences of potential between metals are commonly expressed in the form of a scale known as the Galvanic Series. From the negative (anodic) to the positive (cathodic) ends is as follows for the base metals: MAGNESIUM - ZINC - ALUMINIUM - CADMIUM - IRON/STEEL - LEAD - TIN - NICKEL - BRASS - COPPER, the latter ones being the least active metals. If water flows over adjacent metals an electrical potential difference is generated which can lead to rapid deterioration. This is a fairly common defect arising from the juxtaposition of aluminium or steel and copper in roofing situations. See also chapter on METALS and INCOMPATIBILITY OF MATERIALS.

Water as Fire Retardant

Many of the materials which are accepted as fire retardants in building - e.g. concrete, gypsum plaster - retain this quality because considerable water is entrapped in the hydration process.

WATER IN THE CONSTRUCTION PROCESS

In the construction process considerable quantities of water are used in the `wet trades' (concreting, stone masonry, plastering, bricklaying, blocklaying, drainage) and water free of pollutants is of course desirable.

As water supplies frequently contain significant quantities of minerals, which cause what is known as `hard' water, it is advisable to check with local authorities (water supply, health and building inspectors) to ascertain if any precautions need to be taken with the use of the local supply for building purposes.

Sydney is particularly fortunate, as its supply is drawn from almost uninhabited areas, where the major rock formations are sandstone. This material does not usually contain the calcium causing `hard water'. An attempt to create a lather with water and soap in hard water produces very different results from that experienced with soft water, and the `hardness' is readily noticeable. Water from springs, wells, bores, etc. is often `hard'.

Many areas are not fortunate enough to have soft water supplies. As populations increase and demand for water grows, desalination and re-use of water becomes nearer for many communities. Where this applies, analysis of water may be necessary.

Some US experience with the situation is outlined in Chapter 1, McGuinness & Stein, `Mechanical and Electrical Equipment for Buildings', Wiley,

While the `hardness' ingredients may not be harmful in the normal quantities, sometimes over periods of time, deposits of salts, etc. can build up to create long-term problems. These problems are only now becoming apparent in restoration and maintenance projects on older buildings around Australia. (See `Restoration and Maintenance of Masonry Walls', published by The National Trust, NSW Division.)

The chemicals in the mixing water for cement, mortar or concrete need to be known. Some chemicals, even in small quantities, can be very influential on the setting and strength characteristics expected.

For instance, sugar will completely retard the setting of cement. This is useful to know if faced with an emergency situation, as in the case of a construction collapse.

Structural deterioration of steel beam and adjacent damage to concrete due to rust.

Deep staining of sandstone from rusting wrought iron fence post.

Underground water near the surface must be anticipated on sites adjacent to water bodies.

WATER

Efflorescence in header courses of a brick retaining wall due to salts conveyed from soil by water.

DAMAGING EFFECTS OF WATER
Some of the damaging effects of water on building materials or buildings may be summarised as follows:

Structural
Softening of foundation soils, causing loss of bearing value under footings.
Heaving (expansion) of clay soils, causing uplift.
Overturning and collapse of retaining walls, often due to combination of soil slip and water pressure.
Foundation slip on hillside sites of clay soil. Erosion of soft foundations by running water.
Cracking of stone and brickwork by freeze-thaw effects in cold climates.
Collapse of ceilings due to condensation dripping off metal roofs.
Corrosion of structural members, reducing load capacity.
Decay in timbers.

Chemical and physical
Transfer of soluble salts in porous materials by capillary or surface movement (efflorescence on brickwork is caused this way).
Rising dampness in masonry walls.
Removal of soluble minerals from some materials - e.g. limestone.
Formation of acidic or alkaline compounds with airborne pollutants (aerosols) causing discolouration and corrosion.

Water Penetration of bronze cladding caused disfigurement by lime leached from the encased concrete column.

Corrosion of metals, especially of steel, by formation of iron oxide (rust) in moist situations.
Deterioration on concrete by water containing sulphates.
Condensation causing dampness in internal materials and finishes.
Bubbling of paints due to entrapped moisture in materials. Sagging timber gates, doors, windows, etc. due to variations in moisture content.
Ice accumulation in alpine locations due to wind-driven snow penetration or condensation.

Organic
Fungus growth and decay in poorly ventilated timbers. Development of moulds and mildew on damp surfaces.

Underground Water
Water is sometimes encountered within a few metres below the natural surface, either flowing down an interface between materials or static. The flowing water can cause problems with further excavation and construction procedures and may demand unexpected diversion, drainage, pumps or waterproofing to be installed. Static water usually relates to some adjacent river, swamp, lake or the sea which has saturated the subsoil. The level of such water often varies with seasonal rainfall and tidal influences. Excavation, drainage and foundation requirements are greatly affected by such conditions. The chemical content of ground water can sometimes affect the choice of foundation materials (see FOUNDATIONS).
Steel and timber which are seriously affected by air and water combinations survive much better in fully immersed situations, such as for piles in swamps.

CONCLUSION
In considering the properties of any building material it is essential to investigate the material's reactions with water in all its states.
For external use a permanent material needs to have good resistance to the damaging effects of water.
Materials not so resistant to water effects are often used internally but need to be protected from accidental absorption of water. In many internal situations water-resistant materials should be chosen either as coatings for other materials or the base materials.
Roofing materials need to be compatible with adjacent materials to avoid galvanic reactions. Roofwater in built up areas should generally be collected and carried clear of the building by eaves, gutters and rainwater pipes.

Standards
AS/NZS 1170.3:2003	*Structural design - snow & ice*
AS/NZS 3500.0-1-2-3-4.2003	*Plumbing & Drainage*
AS 3706	*Geotextiles*
AS 3740-1994	*Waterproofing wet areas in residential buildings*
AS/NZS 4347	*Testing DPC's & flashings*

References
Bluescope Lysaght Referee regarding roof & gutter systems

CSIRO Publications
NSB 30	*Building materials in tropics*
NSB 43	*Natural air movement*
NSB 52	*Dampness in buildings*
NSB 59	*Efflorescence*
NSB 61	*Basic facts of condensation*
NSB 78	*Problems with condensation*
NSB 83	*Ventilation in industry*
NSB 76	*Water repellants*
NSB 89	*Sub-surface drainage*
NSB 161	*Building materials in the Pacific*
NSB 166 &167	*Waterproofing basements*
NSB 176A	*Vapour barriers*
BTF 03	*Condensation*
BTF 17	*Plant roots in drains*
BTF 22	*A builders guide to preventing damage to dwellings*
CBPI	*Frost resistance of brickwork in the Australian Alps*

FOUNDATION MATERIALS
*Support for footings; the Earth's crust, rock types, shales, clays, sands,
gravels; conglomerates, alluvial soils, slip areas; investigations.*

Introduction
The safety, stability and permanence of a building depend very largely
upon the crust of the earth immediately beneath it. While many very
populous areas of the world, including large cities, are located in areas
where the earth's crust is geologically unstable it is only the 20th century
that seismic forces on buildings have been studied and this knowledge
applied to building codes. The more conventional knowledge relating to
foundation materials assumes that the earth's crust in its natural forms is
sufficiently stable to support buildings, and the materials that compose
the crust can be broadly categorised into types, with performance
characteristics allotted to them.

It is only this very generalised knowledge which we will deal with here,
recognising the fact that considerable local variations will often occur.

The most important characteristic for a foundation material is its ability
to resist compressive stresses imposed by the footings of a building,
which bear the total structural and imposed loads from the building.

Building codes do not usually concern themselves with other material
characteristics, but some of these can be important in respect of reactions
with certain building materials. This sort of knowledge is usually best
obtained from local building inspectors or testing laboratories familiar
with the area. Acidic qualities in ground water, for instance, can be a
serious problem regarding cement, concrete and steel.

Reactive or expansive clay soils which are affected by variations in
moisture content present special problems, especially for shallow
domestic-type footings.

THE EARTH'S CRUST
The crust of the earth on which we build is incredibly varied, yet all of
geological origin. Variations in soil and rock types, depths of soil, lines
of rock surface, faults, seams, etc. all affect the foundations of major
buildings.

Some cities are built on mud, some on rock, many on clay. Buildings
sometimes have to depend on two adjacent materials of different
characteristics to support them.

A basic understanding of the geological formation of the earth's crust
is useful.

Igneous rocks are the result of volcanic eruption, intrusion or upheaval
bringing molten magma from the earth's core to the surface or crust. In
some cases basaltic intrusions are quite small, while granite areas are more
often associated with mountain ranges, probably formed by geological
pressures forcing the large subterranean granite layers upwards.

Most other rock types are the result of secondary geological action such
as erosion, sedimentation, pressure and heat. Biological content such as
coral, sea shells, etc. in the production of limestone is also important.

The concepts of geological time over which these actions have occurred
are hard to grasp, yet these same actions are still operating today and
become news only when they seriously threaten areas of human settlement
in the form of earthquake, floods, etc.

This slow application of tremendous forces can produce folds in ‚solid'
rock, large cracks or 'faults', displacement of corresponding sedimentary
layers, etc. - all of which can be seen in excavations for man-made
constructions.

Not until the excavations for buildings are made can the exact foundation
conditions be determined, although observation of local conditions and
test bore drillings enable reasonable predictions to be made.

MATERIALS CATEGORIES
Foundation materials for building purposes are frequently categorised
in a simplistic manner such as:

igneous rocks - sedimentary rocks - metamorphic rocks - shales and
clays - gravels and sands - loam made up or filled ground - swampy.

IGNEOUS ROCKS
GRANITES have comparatively coarse crystalline grains and erode
down to a coarse gravelly soil, usually of a khaki colour. In granite country
it is common to find boulders of rock embedded in the eroded gravel, so
that it is often difficult to obtain a consistent foundation material and
level for footings.

Granites can be found in a wide variety of colours from whitish grey to
black, and from pink to red, but in most examples there is a combination
of two or more coloured crystals making up the overall tone.

BASALTS, with their dark fine crystalline formation, erode down to fine
reddish soils and are highly regarded for farming purposes. The outcrops
of basalt are usually in smaller pockets than granite, formed by geological
intrusions, and are now valuable as the quarries for 'blue metal' that is so
much in demand for roadworks and concrete manufacture.

Basalts are usually a black to blue-grey. If brown in colour they are
sometimes called by other names such as 'trachyte'.

Safe bearing values
The safe bearing values for igneous rocks are usually the highest
allowable for building purposes. Local codes allow 4280 to 8560
kilopascals. The highest values are allowed for seam-free conditions
ascertained from site drillings.

Existence of cracks, seams or other materials, etc. reduce these values.

*Typical granite country showing boulders outcropping on a hillside.
A cutting in granite shows the unevenness usually associated with this
rock excavation.*

FOUNDATION MATERIALS

SEDIMENTARY ROCKS

Sedimentary rocks are formed from materials laid down over centuries of time, compressed and adhered together by the action of water, pressure and/or chemical action.

The main types of sedimentary rocks are sandstones, limestones and conglomerates.

Sandstones

These widely occurring rocks usually display layers which sometimes vary in colour, texture and hardness.

The white sandstones are usually the softest, while the creamy yellows, browns and purple colours indicate increasing degrees of hardness often caused by iron-based chemicals combining with the sedimentary materials.

Layers of clay, shale or sand often occur beneath the apparently solid sandstone surface and many surface rocks are in fact large 'floaters', surrounded by softer materials. Whilst these provide adequate foundations for many domestic buildings, larger structures need to have footings taken down to solid foundations.

Sydney city is fortunate in having very sound sandstone under most of its central business district which provides an excellent foundation for tall buildings. This type of stone can also be excavated to vertical faces adjacent to and beneath the level of existing footings without affecting the safety of adjacent structures.

Sandstones can usually be excavated with modern heavy mechanical equipment where there is room to work these monsters, but for vertical faces and trenches, use of pneumatic drills and jackhammers or blasting with explosives is standard practice.

If excavated stone is intended to be used for building purposes the excavation process needs considerable care and skill not usually found in site excavation contractors. Coastal headlands, road and railway cuttings provide excellent places to study the formations of stone and associated soils.

Limestones

Limestone deposits are usually sediments containing evidence of the ancient remains of marine life compressed and hydrated together. As with sandstones, the stratification is often visible and fossilised remains are frequently found in limestones. The texture and colour of the rocks is often similar to sandstone but usually of a finer grain.

Because limes in limestone are soluble in water it is not unusual for limestone country to contain subterranean caves which have been formed over thousands of years, caused by water erosion.

It is from limestones that much of the raw material for cement is obtained. Cement works are often located in areas where limestone is abundant.

For foundation purposes limestones are usually considered as similar to sandstone, but some forms of limestone such as chalks are very soft and unable to support high bearing pressures.

Local building control authorities should be able to indicate suitable bearing values.

Coastal headland of sandstones, shales and clays clearly indicates varying degrees of hardness in the strata.

Heavy machinery excavating solid sandstone to a vertical face for a building basement.

METAMORPHIC ROCKS

Metamorphism is the transformation of rocks into new types by the recrystallisation of their constituents. The weathering processes are not included, as the original rocks can be igneous, sedimentary or metamorphic.

Metamorphism occurs from the effects of heat and/or pressure due to geological conditions.

Shale, for example, develops from layers of clay subjected to pressures and/or heat. If subject to extreme heat the shale can become mica or quartz.

Marbles are usually limestones that have been subjected to heat and/or pressure.

The variety of such rocks is great, and names vary with locations, so that a check for local terminology with Council engineers and building inspectors is always advisable.

Shales

Shales are fine-grained earthy sedimentary and/or metamorphic rocks of thinly layered structure, usually of grey, yellow, green or reddish tones. Considerable variations of colour and density occur in successive layers in some areas.

Shales generally are not as hard as sandstone and the igneous rocks, being composed originally from fine sediments of clay minerals accumulated in water. They are difficult to excavate by hand tools but readily succumb to heavy-powered rippers and can be excavated to almost vertical faces. Exposure to weather leads to erosion and flaking away. Very soft seams sometimes lead to collapse of the harder masses nearer the surface.

As a foundation material shale is usually considered stable and sound with a high load-bearing capacity, especially when protected by topsoil or clay.

CLAYS

In building terminology clays are natural soils that expand and become plastic when wet, retain water to dry out slowly, and are still cohesive when dry.

Clays are found in a range of colours varying from charcoal black, through reds, browns, yellows and white. They can be excavated readily by hand tools and frequently occur as an overlayer to Shales. They are one of the most commonly found materials near the earth's surface, composed of very fine granules of minerals resulting from the weathering of rocks or other geological actions.

Some clays are notoriously treacherous for foundations due to their extremely high degree of expansion and contraction between wet and dry conditions. Local knowledge can usually provide warning if such conditions prevail, often referred to as reactive soils.

A fairly common situation is to find clay up to 1800mm (6 ft) thick overlaying shale. In some sandstone outcrop areas clay is also encountered in pockets both above and below sandstone.

FOUNDATION MATERIALS

Cavities in apparently solid rock can be encountered in excavation.

Sudden changes in sub-surface conditions revealed by site excavation.

The change in plasticity of clay with moisture content makes it a dangerous foundation material. As most clay areas retain a fairly constant moisture content about 1800 (mm (6 ft) below the surface, some form of deep footings are usually advisable for buildings on clay.

In unfamiliar areas it is advisable to seek local advice regarding behaviour of reactive clay soils - and observe walls, local road and railway cuttings and embankments in the area, in order to develop understanding of local conditions.

Excavations in dry or moist clay may readily collapse if subjected to very wet conditions.

SANDS

Sands are composed of fine grains of hard minerals or shells usually washed into water where the hard and heavy grains sediment and are separated from the lighter materials. With the passage of time sand deposits build up and can be retransported by water, wave or wind action.

Because of this tendency to erosion sand is often considered dangerous for foundations, but if properly understood and used it can be very stable and effective, especially in situations where it is possible to reach a level of consistent dampness within, say, 1000 mm of the surface.

The dampness of sand (moisture content) affects its ability to retain a near-vertical face - a fact many learnt as children on a beach using bucket and spade. There is an optimum between 'too wet' and 'too dry' where sand reaches its best, most cohesive state.

Cohesiveness is a desirable quality for building foundations. Saturated sand, however, if well contained so that it does not flow, is a good building foundation.

Because of its porosity sand is sometimes the filter through which water moves under the influence of gravity so that ground water is common at the bottom of a sand bed which overlays clay or rock. Nomads use this knowledge to find water in drought periods.

Sand dunes in waterfront locations, particularly near the ocean, provide very risky foundations. The watertable is usually close to that of the adjacent water mass and can vary greatly with tidal and seasonal conditions. The contours of the dunes can be quickly and dangerously altered by wind and storm action, causing collapse of retaining walls and other structures. Several surf clubs and houses along the sea coast have suffered this fate.

The developing awareness of the waterfront has shown that preservation of these dune locations in near natural conditions is critical in coastal and marine ecology.

GRAVELS

Most gravels are a combination of coarse and fine particles of rock, sand and soils. As such they are like loose conglomerate rocks, retaining some of the characteristics of each component.

Generally gravels are less prone to erosion than are sands, but they may develop some of the plasticity of clay when wet. Stability of gravels is usually good because the proportion of clay particles is not sufficient to create major expansion and contraction problems.

Granite gravels tend to be rough and coarse, while gravels deposited by river action are usually of rounded stones and particles.

CONGLOMERATES

Geological upheavals, glaciation, erosion, etc. all tend to bring together rock pieces, silts, sands and gravels which sometimes become compressed into rock-like masses called conglomerates. Qualities for such materials are unpredictable, but for foundation purposes it would usually be safe to accept the characteristics of the softest component of the mass as a guide, and have tests carried out on samples before designing footings. If such conglomerates have a high clay content they will tend to be unstable when wet; however, some metamorphosed conglomerates can be very hard and stable as foundations.

ALLUVIAL SOILS OR LOAM

The rich alluvial valleys and river flats have attracted human settlements ever since man turned from hunting to farming to sustain his existence.

The recent deposition and periodic flooding of these soils create very soft foundation conditions. The alluvial layers can vary in depth and sometimes deep pile footings are required to minimise risks during floods.

Attempts to stabilise sand dunes with hardy grasses. The buildings can be seriously threatened by storm erosion due to removal of the natural dunes.

Excavations in gravel need gently sloping banks, but contained gravel is a sound foundation.

Excavations in clay are stable when dry but can collapse in wet conditions.

Unretained filling shows effects of water erosion and loose boulders, Excavation is unpredictable.

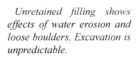

Domestic foundations in this situation are usually kept shallow to avoid the saturated soils related to adjacent water level.

Heavier buildings may need driven pile foundations.

Elevation of the floors above flood levels is usually desirable in such locations, and pile footings readily lend themselves to this arrangement.

The silts usually become extremely soft on the surface when wet and cannot effectively support human or vehicular traffic.

SLIP AREAS

When the natural surface contours of a site are steep (say, in excess of a 30 deg slope) there is sometimes danger of soil `slipping', especially if it is of a clay nature. The greatest danger occurs when the clay becomes wet and loses some of its cohesive quality.

In areas where such conditions are known to occur the local Council usually has some knowledge of the dangers and prepares maps which can be inspected.

Roads and buildings constructed in these locations need to have special geotechnical surveys carried out to ascertain desirable footing types.

Often rock `floaters' occur under the surface and need to be checked carefully before use for foundations.

There is also great danger of surcharge loading on retaining walls, or walls not designed as such becoming retaining walls as a result of slips, and a conservative approach is very necessary.

FOUNDATION MATERIAL INVESTIGATIONS

For specific building sites it is advisable to take test drillings in several locations to ascertain the sub-surface materials and the levels at which they occur.

This investigation is usually carried out by trained personnel with mechanically operated earth augers, mounted on trucks and capable of boring down several metres until a strata of rock or shale is reached. The auger diameter may be small (say, 100 mm) or larger - up to 300 mm. The core is carefully removed and measured, and a report prepared by the firms who hire out such equipment and personnel.

This foundation report gives a fair assessment of likely conditions; but it must always be remembered that rock and ground conditions can change greatly over short distances.

The true nature of the foundations can be determined only by excavation in the final locations for pier holes and other footings.

Standards

AS1289	Methods of testing soils for engineering purposes
AS2121:1979	SAA Earthquake Code.
AS2159:1978	SAA Piling Code.
AS 2758	Aggregates & rack for engineering purposes
AS 2870-1.996	Residential slabs & footings, with supplement
AS 3796:1998	Guidelines on earthworks.
AS 3798-1990	Guidelines on earthworks for commercial & residential developments
AS 4133	(several sections) Methods of testing rocks for engineering purposes
BS1924:1975	Methods of test for stabilised soils.
CP101:1972	Foundations and substructures for non-industrial buildings of not more than four storeys.
CP102:1973	Protection of buildings against water from the ground
CP2004.'1972	Foundations.
NZS 4204P. 1973	Code of practice for foundations for buildings not requiring specific design.
NZS 420SP.'1973	Code of practice for design of foundations for buildings.

Books

Many books have been written on the subject of foundation materials and these are available for more detailed study in the Building Construction, Engineering and Soil Mechanics sections of libraries.

The Design Performance and Repair of Housing Foundations by John Holland - Swinburne College Press, contains valuable case study references to Australian expansive clay soils and their effects.

A. J. Handbook of Building Structure, Section 4, Foundations, gives a general coverage of the topic and leads into detail design for footings.

FOUNDATION MATERIALS

A deep railway excavation in solid shale approximately 100 years old.

CSIRO Publications

BTF 18	*Foundation Maintenance & Footing performance*
BTF 19	*A Builders guide to Preventing damage to dwellings*
BTF 22	*ditto Part 2 Construction Methods*
NSB 2	*Sand gravel rock footings & foundations*
NSB 6 & 9	*Plastic soil considerations & footing design*
NSB 74	*Pier & beam footings*
NSB 113	*Footings & foundation movement*
NSB 155 & 156	*Bearing piles; types & capacity*

Typical sandstone surface outcrops on plateaus around Sydney with varying soil cover.

Rock ledges providing sound foundations on steep hillsides.

PRIMITIVE MATERIALS

A hut constructed from mallee roots, mud and clay caulking, timber and reed thatched roofing.

Round and split logs make the walls of this shed. The roof construction is round log rafters with thatch covering.

Sheet bark, as in this illustration, was commonly used for roofing by pioneer settlers.

These photos by Wesley Stacey reproduced from 'Rude Timber Buildings in Australia' Philip Cox, John Freeland; by courtesy of Augus & Robertson

PRIMITIVE MATERIALS

Adobe; pise; cob; puddled clay; stabilised soil; rubble stonework; limewash;
light timber frames; log cabin, split timber; reed huts, wattle and daub;
thatch; shingles and shakes.

Introduction

Much of this book is devoted to an examination of materials which are widely available in most industrialised societies. Their delivery to the building site is made possible by preliminary stages involving high technology. Modern transport, itself a high-tech industry, carries some products such as glass, marble, ceramics, timber, etc. halfway across the world before they finally reach their destinations.

In this chapter it is intended to consider briefly some of the materials which preceded this situation, are still available and used for housing people not so deeply committed to industrialised processes.

When world population figures and demography are considered, it becomes apparent that this group is more than half of the total. They live in villages and towns on every continent, major and minor islands. In some cases their buildings have changed little over centuries and are generally the product of local materials, labour and skills. The understanding of these has been handed down through the necessity for shelter to be cheaply and locally produced.

The term 'primitive' heads this chapter to distinguish these materials from the high technology products which dominate the cities of the 20th century. However, earth walling particularly has been formally tested, documented and accepted under some building codes and enjoys popularity in residential areas of cities in some countries.

Primitive materials will be considered under two major categories:

(a) geological products

(b) fibrous or botanically developed products

Both are combined in numerous situations and some examples are given.

This chapter differs from the format of most others in that it describes mainly techniques rather than the materials' qualities. The variability of local weather, materials, skills, etc. make predictions quite inconsistent compared with the strictly controlled and standardised materials and components of factory products.

This chapter is included as part of the historical background and in an attempt to remind readers that there are still alternatives to the high technology solutions seen in most towns and cities.

Unfortunately building regulations usually assume certain commonly available materials will be used. For small communities, however, where a very large proportion of the world's population live, it would be wise to retain the older style building technology while materials are available and avoid the horrible shanty towns of manufactured materials visible in many areas. By use of traditional materials and skills and proud owner-builders, the attractive consistency of old established building techniques may still be continued.

GEOLOGICAL PRODUCTS

Adobe or Mud Brick

The city of Jericho, constructed 10,000 years ago, contained three-roomed houses using this method of construction and it is still commonly used in parts of Africa, the Americas, Australia and elsewhere, generally in medium to low rainfall areas.

Local soil or a mixture of approximately 50/50 sand and clay is mixed and moistened with water and puddled (often by treading) until a workable consistency is achieved. It is thrown into moulds on the ground which vary in size, by local practice, but usually larger than bricks and up to concrete block size, say 300 to 400 mm long by 100 to 300 wide.

Moulds are usually made of timber boards of fairly solid construction with handles to simplify removal of the mould from the moulded block on the ground.

Mud blocks in preliminary drying, stacks and walling. The mix for these blocks contained some cement and was effectively done using a rotating cement mixer.

By keeping the mix stiff the moulds can be quickly removed and re-used, especially if first sanded dry before filling.

The blocks are left to dry initially then stacked in open fashion. Walls are built using mud mortar which can have an added portion of lime. The blocks are bonded together in brickwork fashion to produce a wall of minimum 300 mm (12 in) thickness.

Some buildings several storeys high use this material with thicker walls at lower levels.

The resulting walls are excellent heat and sound insulators and merge dramatically with their environments; however, they are subject to slow erosion by moisture or prolonged wind. Extensive research and documentation is available through EBS Bulletin No. 5 ® Earth Wall Construction.

A plaster of mud is often applied and traditionally this has included horse or cow dung, which improves adhesion and protects the structural wall.

Limewash has also been used to protect adobe walls.

A finished adobe house which also employs an earth-covered roof over a waterproof membrane.

PRIMITIVE MATERIALS

Pise or Rammed Earth

Pise buildings have been in use for more than two thousand years and some known buildings surviving today in Europe are several hundred year's old.

This technique uses an earth mixture of approximately 25% clay and 75 % sand or fine gravel. The mix is puddled in a central position (often inside the proposed walled area), then placed and rammed into movable wooden forms. The moisture content is critical to the success of the operation.

The walls are usually 300 mm (12 in) thick, or 450 mm (1 ft 6 in) for two-storey sections. They are placed in forms of up to 3 metres long and 300 mm to 900 mm deep, which are specially designed to permit easy removal and refitting.

Dampcourses can be built in, and protective base courses of brickwork or stone are sometimes used. Openings are readily formed as work progresses.

The work needs protection from rain, especially at the base, and tile cappings, wide eaves overhangs and base veneer courses of brickwork are common details associated with the completed work.

Qualities are similar to those for adobe regarding excellent heat and sound insulation, but the lower clay content renders walls more subject to weathering erosion.

Walls are often mud-plastered or rendered with lime and sand and whitewashed with lime.

Cob

Straw is sometimes added to the traditional earth-and-water mixes to produce walls known as cob. The process is similar to pise but the formwork is omitted and the layers of material are trodden into place and left to dry before another course of material is laid.

This is a slow process and requires trimming of the wall faces and protection from the weather, as work proceeds.

Limewash

Limewash (sometimes called Whitewash) is a lime solution, made up from rock or lump lime and water, in the proportion of one lime to three water (by weight) until `boiling` has ceased. More water is added before use to make a milky wash.

It is applied by brush to the treated wall surfaces in repetitive coatings and allowed to dry out. In the process the lime in solution and/or suspension adheres to the porous earth or masonry surfaces and dries, turning white as it does so. The thin lime skin protects the sub-surface, often softer erodable material. As the successive coatings build up the surface becomes whiter and more waterproof, forming a thin limestone coating.

Where porous bricks, stones or mortars are used, this treatment adds greatly to the waterproofing of the walls. In hot climates the reflective qualities of white surfaces are also an advantage.

In some regions the annual limewashing has become a ritual with the protective purposes probably lost in the past, and all buildings are treated this way, thus maintaining a sparkling fresh appearance which often belies the real age of the buildings.

Puddled Clay

This material has been widely used for floors on the ground, for hearths, ovens, and for raised floors and roofs where it is beaten on to a timber and brushwood framework.

Floors should be protected from water entry, usually by a raised threshold. Some Mediterranean houses feature split-level floors of this material with the higher portion approximately 450 mm (18 in) above the lower portion, allowing it to be used as a seat and raised sleeping area.

Corners are rounded.

The resultant floors have a pleasant feel underfoot, especially when a proportion of sand is included in the mix.

Stabilised Soil

Techniques of improving the weathering qualities of soils and reducing the liability of volume change involve the introduction of some high technology products. Where these are available cheaply they may be worth consideration.

Cement is the major stabiliser used widely for road construction and can be adapted for building purposes. The proportion of cement to earth needed is in the vicinity of 1 to 10, but smaller proportions can have a significant effect. For economy's sake the treated areas need to be minimised and restricted to external surfaces or pavements.

One type of movable wooden formwork for pise wall construction.

Thirty-year-old pise wall samples showing a variety of soil mixes at the Experimental Building Station, North Ryde.

Bagged lime can also be used similarly.

Mixing techniques for these powder-type materials usually involve spreading and raking over and into the dry earth. Thorough mixing is difficult if the earth is damp. Use of rotary-type cement mixers is usually not effective because of the tendency of clay to ball up into lumps. Water is sprayed over the earth and then it is smoothed off or rammed in place.

Rubble Stonework

Use of local stones is common in many areas where stones small enough to be manhandled are readily available near the surface of the ground. These can be laid up into rough stone walls with a minimum of cutting and fitting, sometimes using smaller stones or chips as packing. These walls are known as rubble stonework.

Rubble walls need to be fairly thick (350 + mm) to be stable enough to support roofs and be weatherproof. The use of lime mortar improves the stability, weather-resisting properties and speed of construction.

The natural shapes of rocks and stones available greatly affect the appearance of the finished wall. Simple tools such as hammer and bolster can assist in cutting and shaping to produce neat fitting and stable results.

The major problems with rubble walls relate to supporting work over door and window openings. Good quality timber has often been used for this purpose as large stones needed to do this in masonry may not be available or are too heavy to handle readily.

The chapter on 'Stones' should be referred to also regarding material characteristics, but for most uses well-constructed stonework will outlast its builders.

Dry jointed rubble stone wall.

Rubble stonework laid with mortar joints.

FIBROUS MATERIALS

In building with these materials the first consideration is 'What grows locally that can be used?' The range is considerable. In some cases there are numerous choices; in others, little or none at all.

All natural fibrous building products are botanical in origin and therefore are subject to decay. The resultant limited life span of buildings in some locations is accepted, as in Fiji, where huts are burnt ceremoniously after a replacement has been constructed.

If well constructed, however, these materials are adequate for their local conditions and may well outlast several generations of a family to whom they have provided shelter and home comforts.

It is not unusual for some items such as thatched roofs to be completely removed and replaced occasionally, or for newer material to be fixed over the original.

Some traditional Chinese and Japanese monuments of timber have been rebuilt several times over the centuries of their existence, without changing the traditional techniques or appearance.

Light Timber Frames

A framework of timber to support wall and roof coverings is a basic form of construction and probably the first used by man in his efforts to provide domestic shelter of any permanence.

With very primitive axes materials can be procured by cutting light branches from trees, shrubs, bamboo or other plants. To join these members together into a stable structure various techniques have developed which remain in use in many areas today.

Lashing members together by use of fine or shredded vines or split bark enabled secure joints to be made where two members crossed each other. Sometimes at the crossings members were notched out to provide a secure seating.

Infill between larger members was often provided by weaving light members at right angles to each other as a base for the final weather shield of more vegetable material or clay (wattle and daub).

In some cases the lashings become a continuous method of fixing the major and minor members together into a lashed and woven system of great skill and decorative potential.

Examples of these techniques are still demonstrated by natives of Africa, Polynesia, New Guinea, etc.

Bamboo is often used in this type of construction.

Light round timber framed and lashed roof structure showing also underside of palm thatched roofing.

The Log Cabin

Heavier construction more suited to European weather conditions and trees was developed about 1000 BC in the form of the log cabin.

This form used full logs of straight forest trees trimmed of branches and halved at ends where the members crossed. With plentiful timber from natural forests this was easy. Developments at about AD 1000 involved splitting the logs to reduce the timber needed. Shaping the horizontal log junction to minimise water penetration and draughts was a Scandinavian improvement.

The crude joints were often plugged with mud to minimise weather penetration. Windows were not used or they were kept to a minimum, and the usual one-room building remained stable by virtue of the interlocking corner members and the building's own deadweight.

Roofs were often of bark laid over smaller logs and held down by tree branches laid on top of the roof.

Where straight softwood timbers abound this method is still useful, as was demonstrated in North America in the great westward march of the latter part of the 19th century.

Modified log cabin construction.

Round timber framing as used in many South Pacific areas.

Split Timber Construction

With the ability of many softwood timbers to be split easily, it is possible to use the splitting technique to minimise material by getting two or more members from each log or branch.

Bark can also be shredded by splitting into lengths usable as lashings, or by flattening out to use as roofing or wall covering. This technique allows framing members to be developed that are of reasonably uniform sizes which can be connected by steel nails (a concession to modern technology). Timber dowels were often used where nails were not available.

Reed Huts

The use of light, flexible reeds and palm leaves is referred to in Egyptian records. They are often woven together into a form of mat. Modern usage persists in the Japanese Tatami mat and in Polynesian huts.

The Egyptians sometimes plastered the reed walls with mud.

Split slab hut over 100 years old in which the original shingle roof has been covered with corrugated iron.

Tropical palm thatched roofing with split bamboo wall cladding.

Wattle and Daub

This term applies generally to a method of infilling between a main timber frame with light split saplings and applying a coating of mud on one or both sides. The mud was often mixed with cow dung, for improved adhesion.

This method is very similar to that used for the flat roofs of many houses in Mediterranean countries where rainfall is low. The saplings and brushwood structure are covered with the mud and dung mix, which is beaten into place. This roof is a useful insulator and provides usable outdoor space on hot nights, but is not reliable against rain penetration.

Thatched Roofs

Roofs able to resist rain penetration have always been a problem with buildings. In areas where snow also falls this adds another problem for roof construction.

As a result of these factors it is very common for roofs in many locations to be pitched more steeply than is necessary merely to drain away the water. A deep build-up of snow adds weight which has to be supported by both roof and wall structures.

For both water and snow to be easily shed from a roof the roof needs to be sloped generously. When this is done the natural flow of water due to gravity can be picked up and carried clear of the walls by a multi-layered system of materials not individually capable of such protection.

Such is the system called 'thatching'. It can be done using numerous types of vegetable products. In tropical areas it is usually palm leaves or matted palm fronds. In cooler climates thatching is in the form of sheaves of grass, bundles of reeds, straw, or heather, laid on a sloping timber framework, usually of round or split members.

Where grasses, etc. are used, all seeds need to be threshed out to reduce risk of germination and attack by rodents. The dried-out sheaves or bundles are laid carefully from the bottom of the roof upwards and tied in place to the roof framing. The resulting roof covering is approximately 300 mm thick and provides a very good insulation against heat gain or loss, and will give a life of up to one hundred years.

In the tropical and warm climates examples of thatching, the palm frond-type materials are flatter and broader, resulting in a thinner overall depth of roof cover, but several layers are used to ensure satisfactory performance.

In all types of thatching the big danger is fire, as the roofing becomes very dry at times. Secondary problems are the incidence of pests and vermin, using the roof as their home also, with unfortunate results in the spaces below.

Shingles and Shakes

Shingles generally refer to pieces of timber, rectangular in shape, used for covering pitched roofs or timber-framed walling. They are usually nailed to underlying timber battens which, in turn, are fixed to the main framework.

Each shingle is laid with two-thirds of its length covered by the next members above, which are offset sideways half the width to provide cover for the joints between adjacent shingles.

Shingles are usually made from selected species of timbers which are known to be unlikely to split when cut down to approximately 10 mm thickness and exposed to weathering.

Western red cedar from America is frequently used, but many other species have proved suitable and durable. Casuarina provided roofs for many Australian colonial period houses.

Shingles can be sawn or split, the latter giving a more irregular appearance. In. North America these split shingles are often referred to as 'shakes'.

In most primitive situations the split shingles would be used, but mass-produced nails, preferably galvanized clouts, would be the recommended fixing method, for cost effectiveness and efficiency in use of labour.

Similar roof coverings of this type have numerous variations and local details where they are still commonly used, as in some Eastern European villages. The natural movements of the timber, responding to dry or moist conditions, have been exploited to allow additional summer ventilation to the roof, yet expand to seal it adequately in winter.

As with all thin timber member construction, the great danger is fire. It was this factor which virtually eliminated shingles from

Round pole framing with some rectangular members probably split and adzed to shape.

use on Australian buildings once more fire-resistant materials were available. However, the light weight of individual pieces and the complete roof make shingles an attractive proposition in some locations. They are still widely used in North America.

Summary

With all of these primitive materials, the quality of the finished product is very dependent on the local materials and skills available. Few tools are needed.

The skills are readily acquired and a little experimental work on a small scale will indicate preferred materials and proportions. The whole nature of the handwork is rough and extreme accuracy is not important, as it is in high-technology building. The resultant textures contrast strongly with the hardness of modern materials and methods.

Many individuals and communities in industrialised societies are now investigating and using some of these methods to gain the personal satisfaction of building for their own needs.

Traditional knowledge of building has been lost in some areas where a return to the old techniques could well reduce dependence on cash to produce shelter using manufactured products.

In many cases the striking evidence of how well these techniques blend with their landscape shows their compatibility with current environmental movements and philosophy.

The disadvantages are largely associated with the limited life of resulting buildings, especially with fire risks; the industrialised society's expectation of high-quality, low-maintenance finishes; the desire to be able to standardise workmanship and products for contractual and regulatory purposes.

References

The Macmillan Encyclopaedia of Architecture & Technological Change-Building Materials
Rudofsky- Architecture without Architects. Academy Editions 1964
Freeland Cox Stacey- Rude Timber Buildings in Australia
Guidoni- Primitive Architecture

A roof of split timber shingles or 'shakes' on a small reconstructed pioneer schoolhouse.

CSIRO Publications

NSB 13	*Earth wall construction*
NSB 18	*Pise construction*
NSB 30	*Building Materials in the tropics*
NSB 161	*Building Materials in the Pacific*
BTF 06	*Earth Wall Construction*
BTF 07	*Pise (rammed earth) consrtuction*
BTF 21	*Straw Bale Construction*

TIMBER

Elevated floor flaming using round timbers. Wall covering uses split bamboo sheeting.

Hardwood logs stacked at the mill awaiting the saws.

A short length of log being sawn.

TIMBER

Balks and planks during mill preparation.

Deterioration of thin sawn timber boards without paint protection or edge lapping. Weatherboards (overlapping edges) probably more recently applied.

An external timber deck and stairs in which the choice of timber species can be critical for safety and long life.

TIMBERS & BUILDING BOARDS

Joining; softwoods; hardwoods; strength; grading-sawing variations; sapwood; heartwood; species; terms;
chemical constituents; hazards; preservatives; variations; defects; structural timbers; laminated timber; joinery
timber; references; characteristics summary; building boards.

Introduction

The great asset of timber is its lightness for strength available, combined with the ease of cutting, fitting and joining it with simple hand tools.

The visual and tactile qualities of the material have also been appreciated, especially for furniture and internal finishes where the variations of natural grains and figurings can be seen and felt.

As the most easily worked and widely available material, timber probably has been the longest commonly used of all building materials. The reason it does not feature as notably as brick and stone in very old buildings is a reflection of its inherent organic characteristics, susceptibility to fire, decay and insect attack; however, many details and forms in stone buildings are indications of details that must have been influenced by timber antecedents.

Timber in its light forms as saplings and twigs needs little in the way of tools. The heavier forms were acquired from trees felled by primitive axes, then split into useful sized members and adzed roughly to shape. Carpenter's tools to work timbers are known to have existed thousands of years BC, and until the advent of steel bodies and electrically powered hand tools in the 20th Century these changed very little.

Joining timbers

The uses of timber have been closely related to the means developed to join the timbers together. In many primitive tropical societies where vines are plentiful, the use of vines to lash and tie members is still practised. This was the first step beyond literally resting a horizontal member in the fork of a tree branch to help develop a primitive shelter. Lashings are probably the only timber jointing methods that do not need some carpentry tools.

Working timber to make close-fitting joints requires cutting tools such as saws, chisels, etc. that developed with bronze's use as a hard durable metal.

Holes were probably first cut by using red hot metal to burn through timbers. Dowels of timber were commonly, used to hold two members in place, and these still are used in some situations.

Water-powered sawmills developed in northern Europe in the 14th to 16th centuries and the export of timbers began. Hand-wrought metal spikes or nails have been found in many old roofs and timber frames. Wrought iron plates and threaded bolts and nuts became commonly associated with timber trusses in the 19th century.

Mechanically produced nails developed around 1800. In association with power-driven saws, smaller sized sawn timbers and the resulting light timber stud frame (developed in USA in 1833), nails revolutionised the jointing technology of carpentry. The construction of domestic, wall, floor and roof frames by this new technique has since dominated the industrialised world.

Further developments combining the versatility and mass production of steel connectors with the ease of site working in timber have led to gang nails, framing anchors, etc. which aim to retain timber's competitive position in the domestic building market.

Carpenter's glue was traditionally made of boiled-down animal materials and served joiners well for internal work but was not reliable in external situations.

Casein glues came into use about 1930, but during World War II waterproof glues were developed which made waterproof plywood possible and initiated the tremendous variety of the chemically-based glues and coatings now available. On site joints as strong as the components can now be produced if these joints are properly prepared.

SOFTWOODS AND HARDWOODS

As applied in the building industry the terms 'softwood' and 'hardwood' relate to the relative ease of working a species of timber with carpentry tools.

The softwoods tend to be lighter than hardwoods and generally not as strong for structural purposes.

Obviously there is no clear line of demarcation. The terminology is further confused by the botanical distinction that applies the term 'hardwood' to wood that contains pores or vessels in its makeup, whereas softwoods do not.

Botanically the conifer species (cone-bearing trees of which the pines are the best known) are all softwoods while the broad-leafed trees like the eucalypts and most tropical and sub-tropical rainforest trees are hardwoods, despite the low density of some of them.

The density of timbers can vary from balsa, at 160 kg/m3 (or cu m), to 1200 kg/cu m for some Australian hardwoods. See attached table of some popular species and their densities on page 26.

STRENGTH OF TIMBER

The higher the density of timber generally the greater is its mechanical strength. The amount of woody material in a given volume will largely determine the load it is capable of supporting.

In compression and tension, timber is strongest in the direction of growth (i.e. along the grain) and crushes much more readily across the grain.

Stress Grading

Timbers can be graded for strength by 'stress' grading which is done either visually or mechanically to indicate primarily the basic working stress in bending for engineering design purposes. This grade is usually stamped on the sawn timber before leaving the saw mill in a form such as F7, indicating that the basic working stress in bending should not exceed 7Mpa.

Local Standard Codes and authorities usually supply details of the symbols and stress applicable.

A well-maintained timber framed and sheeted house with decorative timber details built on a stone base about 1900.

Visual Grading

This involves the separation of individual pieces of timber into categories for particular end uses. The grade is determined by the number, size, type and position of knots, gum veins, shakes or other visible characteristics that affect its strength and/or appearance.

Since no two pieces of timber are exactly alike complete uniformity within a grade is unrealistic.

Local terminology and samples need to be examined in order to understand the relevant gradings.

Sawing Variations

The way in which a piece of timber is cut from the log affects its appearance and performance regarding shrinkage. This depends on the sawing pattern adopted at the mill.

Books on carpentry and joinery should be consulted for detailed explanations of these variations.

Preparations for Marketing Timber

A critical factor affecting the performance of most timbers is the method and care taken with `seasoning', the process of drying out the sap and moisture from the felled timber to reach the moisture content desirable for the locality and purpose intended. This can vary with different species, the size and nature of the log or sawn member. For furniture the timbers need to be carefully seasoned and some high quality manufacturers keep their sawn timber stock for many years to minimise shrinkage in their finished products. For concealed house framing eucalypt hardwoods are often used `unseasoned' because of easier workability, but kiln dried grades are needed for quality work which is to be visible in the finished job. Major sawmillers and Forestry Authorities have information available.

For many softwoods including the now widely used Pinus and New Zealand indigenous species the whole log can be gainfully included as usable timber. The layer of cells actively carrying nutrient to the tree is nearest to the outer perimeter or bark, and is called sapwood. For indigenous Australian hardwoods this is usually discarded, and only the heartwood used commercially. Structural timbers are now graded for strength, the higher the `F' grading number the stronger it is in resisting stresses.

To maximise resources many timbers used in construction are now chemically treated to reduce susceptibility to decay and attack from fungus and insects. This technology has allowed fast growing softwoods to be more widely used both externally and internally. As the processes involve toxic chemicals they need to be strictly controlled. The New Zealand Timber Preservation Authority and Australian Forestry authorities have publications about these processes which are vital to the renewable softwoods timber industry.

Sawn scantlings assembled into a stud flame.

Selecting Species

Where visual qualities are important in the finished product it is necessary to identify accurately any particular species of timber in order to achieve the required effect, as colour, figure and grain vary greatly between species.

Costs and availability of species can also vary and need to be considered in making the selection.

When strength is the major consideration, stress-graded timbers of mixed varieties may be satisfactory and cheaper than selected timber of one species only.

For durability in certain difficult locations such as unpainted external work subject to wet and dry conditions, the species is of great importance. Local examples as built should be checked for performance, availability and cost.

It is only by local knowledge of timbers that reliable selections can be made for a given situation. Reference should be made to local forestry authorities or timber merchants' association.

Sawn Timber Terms

Some commonly encountered terms used in the timber industry need explanation. They are purposefully not in alphabetical order.

Log - the felled tree trunk, stripped of branches, ready for transport to the sawmill.

Balk - a large sawn timber member approximately square in section and 300 mm to 450 mm on each side.

Planks - formed by sawing the balk lengthwise into sections of, say, 200 mm wide x 50 to 75 mm thick.

Scantlings - structural members such as bearers, joists, plates, etc. usually referred to by their nominal size.

Sawn boards - usually 100 mm to 200 mm wide and 20 mm to 25 mm thick and rough off the saw.

Dressed boards - as above for nominal size but machine planed or `dressed' to produce a smooth face. DAR = dressed all around.

Thicknessed - refers to members dressed on two faces only to a consistent dimension. With structural members it is usually the maximum dimension which is the consistent one.

Battens - timber strips from 25 mm to 100 mm wide and up to 25 mm thick. In roof construction used to support tiles, slates, etc. over short spans. Also used as fixings for internal linings.

Nominal size - is less than actual size due to saw cuts, shrinkage, etc. It relates to the centre line of the saw cut.

Carpentry - is directed toward the timberwork necessary for support, division, or connection, with the object of giving firmness and stability to a structure. It is an ancient art which has been practised for thousands of years.

The timber framework of a building is carpenter's work and therefore `carpentry'.

Joinery - provides the additional elements aimed toward comfort, convenience and ornament. It requires great care and skill in making joints and fitting elements together so that the finished product can be viewed closely.

Timbers used for joinery must be of good quality, well seasoned and selected with understanding of the characteristics and behaviour of timber species. The joiner needs a detailed knowledge of geometry and the means of setting out and cutting timbers to fit exactly.

This level of joinery developed greatly during the last 500 years and probably reached its peak in the 19th century and early 20th century, when timbers were still plentiful and craftsmen were highly trained.

Timber doors, windows, furniture and fittings were traditionally joinery work. Much of this work is now factory produced but is still referred to as joinery.

Cabinet making has largely superseded joinery work because much of the shaping processes are now executed by machines. The elements are frequently timber and timber products such as plywood, chipboard, etc. and the assembly of components is done by comparatively unskilled or semi-skilled labour and machinery.

Seasoning - All timbers contain sap (water) when felled, which needs to be reduced in quantity to achieve optimum working strength and dimensional stability for use in building. This process of drying is called `seasoning' and can be done by natural air drying, which is fairly slow, or else accelerated by kiln drying, which can be done in half of the time or even less, under controlled conditions.

TIMBERS

Domestic trussed roof frame using gangnail joints.

Detail showing grain of Douglas Fir framing and assembly
of studs, plates and joists with chipboard flooring.

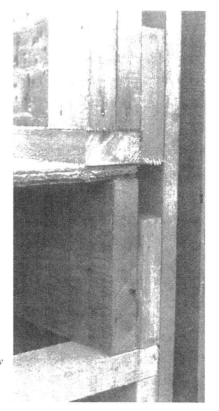

The objective usually is to reduce the moisture content to a percentage
level which is appropriate for the conditions of use. This percentage
varies with different climatic and use conditions from 8 % to 20%, but
12% is a commonly used standard.

In the seasoning process the longitudinal dimensions scarcely change,
but the cross-sectional measurements of the log, balk or plank do reduce
measurably, and sometimes unevenly, depending on the way in which
the balk was cut into planks and other irregularities that can occur in the
growth of trees.

The greatest shrinkage occurs in the direction tangential to the
circumference of the tree trunk. Radial shrinkage is only half of the
tangential shrinkage.

Slight changes of dimension will continue to occur with atmospheric
variations and seasonal conditions. Such movements can cause windows
and doors to jam, wedged joinery joints to become loose, etc. These
movements seldom affect the structural timbers unless directly exposed
to sun and rain.

Some hardwood species are used for structural purposes in a partially
seasoned condition, because when fully seasoned they become so hard it
is extremely difficult to drive nails into them or to drill them.

Wood turning - This term refers to the use of wood turning lathes to
produce circular-shaped members as often seen in stair balustrades, etc.
A great variety of decorative effects is possible through this technique.
Not all timbers are suited to the process, however, and local advice from
craftsmen is necessary for good results.

CHEMICAL CONSTITUENTS

The chemical constituents of timber are mainly: cellulose 45 to 60%,
hemicellulose 15% to 20%, lignin 25 % to 35%, plus resins, oils, tannin,
alkaloids, etc. - all varying according to species. The lignin is the glue
holding fibres together, and a high lignin content produces a high
compressive-strength timber.

HAZARDS TO TIMBER

Termites occur in many of the tropical and temperate regions of the
world. They are often referred to in Australia as 'white ants'. There
are many species. Colonies of these insects can completely destroy the
structural strength of members by eating the timber from the inside.

Special precautions must be taken in termite-prone areas to prevent
the insects entering the timbers from the ground, and to discourage their
presence by clearing away all roots, stumps, timber off-cuts, etc. from
beneath a building and its immediately adjacent areas; by treating timbers
under the floor boards, and/or by poisoning the ground.

Some species of timber such as Australian cypress pine possess natural
chemicals which give them resistance to termite attack. Undesirable
environmental effects have now led to ground poisoning being banned in

Australia. Other more expensive methods such as fine stainless steel
mesh collar barriers have been developed to protect from termite intrusion
through pipe holes in floor slabs.

Part of a termite-damaged structural member. This often remains
concealed behind a thin outer shell of timber.

TIMBERS

Heavy timber construction in which the hardwood columns and beams are reasonably fire resistant.

Fire-damaged building showing how charcoal forms on timber members. Note how light timber laths in wall have been protected by the plaster applied to them.

Fire can completely destroy most timbers if conditions are favourable, and this is certainly the greatest hazard to buildings of timber. However, some timbers, particularly in heavy structural sections (say, at least 100 x 75 mm or larger), are resistant to the fire once the surface has been charred.

Many of the Australian hardwoods have these characteristics and, in fact, have proved to be far more fire resistant in buildings than steel. Nevertheless, all timbers do burn readily if temperatures remain high enough; therefore timber buildings are consequently not classified as fire resistant.

Dry rot is a commonly misused term and should be called decay. The true dry rot occurs only in timber which has an excessive moisture content - more than 20%. This is the result of fungus attack, usually brought about by inadequate ventilation, in which the fibres of the timber are reduced to a dry, powdery dust. The true dry rot of Europe is very seldom found in Australia.

Wet rot or decay is also a fungus disease caused by excessive and continuous periods of dampness that result in decomposition of the fibres. It can occur in external timbers which are periodically wet and dry - e.g. the foot of timber posts in contact with the ground or pavement areas.

Any localised softening of timber indicates some form of decay has developed.

Marine conditions - If fully immersed in water or wet conditions many timbers remain stable for long periods and have been successfully used for footings under major buildings - e.g. Winchester Cathedral in the United Kingdom, and piled footings, supporting Venice and other cities.

However, in marine conditions various organisms attack the timber, especially in the changing wet and dry conditions between high and low water marks. See FCNSW paper, Marine borers and the breakdown of timber in the sea - Keirle 1974.

Some timbers are naturally resistant to these attacks, notably

Australian turpentine, which has provided effective piles for more than 150 years and has been widely exported.

Borers - two main types of borer are common in Australia. They are the Lyctus (powder pest) and Anobium (furniture) beetle. The former attacks only the sapwood of certain hardwoods, and local laws in some States control the use of such Lyctus-susceptible material. Anobium attacks softwoods and some hardwoods. Simple and effective chemical treatments protect susceptible timber against these insects. Neglect can lead to great loss of strength and disfigurement due to the small, round holes left in the timber. Lyctus borer activity usually is indicated by the powdery residue left by the insects.

Weathering - When exposed to weather, timber swells and shrinks with moisture variations which can cause cracks in the surface of the timber and further deep penetration of moisture and subsequent decay. Painting with pigmented external paints will normally prevent this, but regular maintenance is necessary, particularly at joints and flat surfaces where timbers cross in exposed position, as in pergolas.

Exposed joint surfaces should be painted before fixing together. Timber stains do not usually give as long life as paints and need more frequent applications for adequate protection. A few timbers can survive unpainted and unharmed for long periods. Some Australian hardwoods are suited to this for engineering-type situations where members are at least 50 mm thick. Thinner members often split and crack.

Many Australian eucalypt timbers create a strong brown stain when wet by rain, and this can seriously disfigure materials below such exposed timbers. The stain is difficult to remove, but it can be painted over.

Where clear stains are being used externally with timbers, the fixings (nails, etc.) can cause rust marks which disfigure the appearance.

Heavy hardwood planks can give long service in severe conditions without painting.

Floorboards and framing showing clear evidence of dry rot, a form of decay due to fungus.

A termite access tunnel constructed to bypass a metal antcap and reach structural timbers.

Timber preservatives

Surface coatings are often used to minimise the hazards to timber.

The most common method has been to paint timber with coaltar creosote. This, however, is almost black in colour and highly flammable. It is an effective preservative but does not penetrate far into the timber. It is in the short term reasonably effective against termites.

Some soluble salt treatments are available that can be used and later painted over, but again the degree of penetration is very limited.

Pressure impregnation processes are more efficient and can achieve maximum absorption. They have to be carried out in properly equipped closed cylinders and often use toxic chemicals that leave the timber slightly stained. Some preservatives can also in-corporate flame retarders or water repellers.

Suppliers of these preservatives should always be consulted for advice regarding the most suitable treatments for particular timbers and situations. Examples of the resulting products' weathering capability should be examined to verify the often exaggerated claims of durability.

Variations in timbers

There is tremendous variation in the types of timbers which are grown in different regions and the consequent common usage and availability of timbers.

Due to man's constant plundering of the natural forests many areas are very short of timber for building and so many timbers have to be imported, usually by sea.

Australia, especially the Sydney region, has been importing since the 19th century, largely because of scarcity of suitable softwoods. The few natural softwoods suitable were largely cut out, or cleared and burnt in the rush to create grazing lands from the natural forest areas. This took less than 50 years (1880 to 1930).

The variations in locally available timbers combined with the other variables to which timbers are subject, make it one of the most difficult materials to understand unless one is constantly in touch with it. Certainly, classification systems seem complex and vague compared with those of factory-made products, and timber names vary from place to place.

Consequently, the only practical way to become aware of the inherent qualities of timbers is to handle the materials and preferably use some tools to cut and fit pieces together, and to recognise general characteristics.

Density is one of the most obviously variable qualities and is usually related to structural strength. The density of timber varies greatly from 500 kg/m3 to 1300 kg/m3.

The more dense timbers usually have good compressive and tensile strength. The softer light timbers are suitable for many normal situations but usually have inferior structural qualities and do not hold nails or screws as securely as do the denser timbers.

DEFECTS IN TIMBER

Knots, produced by branches joining into the main tree trunk, cause defects which can reduce the effective structural strength of a member, and in some cases the heartwood of knots can fall out of the board. This heartwood is usually a darker colour than is the main body of the timber. For timber cladding, lining boards, etc., however, knots are often acceptable and add to the natural appearance of the product.

Shakes are defects caused when a tree is severely jarred in felling, causing a longitudinal separation between adjoining layers of wood. (This term is not to be confused with the term 'roofing shakes' which describes purposefully split tree products which have been used as roof claddings since early days.)

Gum veins in some species leave noticeable cavities along the grain similar to shakes.

All of the defects described here lead to a reduction of the anticipated structural strength of members. In low-stress situations such as stud walls, however, they can often be tolerated; but such defects should not be tolerated in beams or other major structural members likely to be subject to high stresses.

TIMBERS
Approximate Densities of Some Commercial Timbers (Seasoned)

Common Name	Density Kg/m3	Common Name	Density Kg/m3
Ash, Alpine	620	Maple, Queensland	580-640
Beech, European	700	Meranti red (Indonesia)	400-720
Beech, white	500	white	550
Blackbutt	900	Messtnate	780
Box, brush	900	Oak, silky	550-830
Cedar, red Australian	420	Pine -	680-710
Cedar, western red (USA)	370	cypress	
Douglas fir (Oregon)	530	hoop	530
Gidgee	1250	radiata	580
Gum, grey mountain	880	white New Zealand	450
Gum, spotted	950	Rosewood - New Guinea	650
Ironbarks	1100-1140	Stringybark	770-900
Jarrah	820	Tallowwood	990
Karri	900	Turpentine	930
Kauri - New Zealand	560	Walnut, Queensland	690
Mahogany -			
Philippine red dark	630		
Red eucalypt	950		
White	1000		

Above table is an abridgement from `A Guide to the Density of Commercial Timbers', by K. R. Bootle 1981.

TIMBER SPECIES
No attempt will be made here to enumerate individual species and their characteristics as this is best done by local reference to suppliers, publications by Timber Development Associations, etc. Some publications that are useful for this purpose and for detail diagrams regarding timbers, as well as some relevant Codes of Practice, are listed at the end of this chapter. Studies should be made on examples with several years exposure to similar conditions before selecting a species of timber for external use.

STRUCTURAL TIMBERS
As structural timbers generally are used in long, rectangular pieces of 35 mm thickness or greater, they need to:
(a) be able to hold their shape well even if not fully seasoned;
(b) hold nails near the ends without splitting; and
(c) be free of knots or defects affecting their structural continuity.
The stress grading system now commonly used for such timbers does give some indication of these qualities, but each species can vary quite markedly regarding the first two requirements. Also, because of such characteristics as splinteriness and density, some timbers become unpopular with tradesmen.

For these reasons eucalypt hardwoods are seldom preferred by Sydney carpenters if pine or Douglas Fir is available. However, to specify such imports in country areas near local sawmills producing eucalypt hardwoods can be politically foolish. Consequently, the structural timbers chosen are not always the most technically suitable but often are the most convenient available unless reasons exist to restrict the choice because of likely stresses in the designed members. For domestic uses this is seldom critical.

The standard sizes that have become common almost universally for domestic timbers (100 x 50; 75 x 37, etc.) are as much a result of the sizes needed to do the necessary joining and nailing as they are of the sizes needed to resist bending and compression stresses.

Sawn timbers are readily available from local suppliers in various lengths, but long pieces, say in excess of 5m, may have to be ordered in advance.

Warehouse buildings more than 50 years of age frequently display excellent quality structural timbers and techniques which would be almost impossible to duplicate today. The timber quality and sizes used then are often unobtainable now. Today these heavy members could be reproduced if necessary by making up laminated members, or by reuse of members from demolished buildings.

LAMINATED TIMBER
Laminated timber is an increasingly popular application of timber for structural, decorative and utility situations. Timbers approximately 25-37 mm thick are usually selected and dressed smooth, then placed together with adjacent faces glued and subjected to pressure. In the process bent shapes can be achieved to follow desired forms and the resulting shaped

components have great stability and structural strength.

Because of the laminating the glueing overcomes inherent weaknesses or defects in each laminated member. The allowable structural strength for such members often exceeds that of the parent timber, and members can be built up to any required size.

Some manufacturers have specialised in this process and can produce structural members at costs competitive with alternative materials.

Specialised shapes also are available.

Reference should be made to local firms engaged in laminated timber construction.

One of the useful characteristics of laminated timber members is their resistance to fire, as the surface usually chars then resists further burning. Some structural members have been reused after fire damage without serious loss of strength.

Laminated members are also useful in corrosive industrial atmospheres which would quickly attack metals - e.g. salt warehouses, acid bath areas, galvanising works

Their pleasing appearance and stability has been their great asset in high quality work such as public building interiors - e.g. the High Court building, Canberra, Australian Capital Territory, 1980.

Laminated timber can be as useful decoratively as for practical purposes.

JOINERY TIMBERS

The traditional uses of timber joinery have undergone major changes in the last 50 years. Timber-framed panelling, doors, cupboards and fittings now frequently incorporate plywood, veneers and coreboards of various types. There is still a call for quality joinery in good class work, and some elements such as moulded timber handrails to stairs are not surpassed by substitute materials.

For good joinery work the timbers usually need to be of a closegrained species which do not readily split, can be brought to a fine, smooth finish and take stain, polish or other clear coatings. If the timber also has a slightly decorative figure, then it is a very desirable joinery material.

Timbers which incorporate these qualities and are available in quantity command high prices and are exported all over the world for furniture and fittings.

Because English oak was one such timber, with a distinctive net-like figure, trees producing similar figured timbers in other countries were often called oak trees although they were of entirely different species. Thus we have local examples of she oak, swamp oak (casuarinas), silky oak (Cardwellia sublimas), and Tasmanian oak (a eucalypt).

Similar confusion in names also exists with Queensland maple and rose maple, which are not related to the Northern Hemisphere maples.

It is therefore necessary for local usage of timbers to be studied in order to become familiar with the common names used and the related characteristics.

TREATED PINE

Pine logs and scantlings from plantation forests, treated to preserve against decay and termites are becoming increasingly used in structural and landscaping work. The preservative used is a copper compound impregnated into the timber under pressure. This gives these products a distinctive green hue, and also makes them dangerous if burnt, as they then give off poisonous fumes. These timbers appear to be prone to splitting when exposed to the weather, and are not as strong as Australian hardwoods, but will resist termites when in contact with the ground. Surface treatments and metallic fixings need to be carefully selected to be compatible with the preserving chemicals.

Elaborately detailed Australian cedar joinery from the 19th century.

SUMMARY OF CHARACTERISTICS OF TIMBERS

Resistance to structural stresses

Considerable variation between species and between samples within species due to varying growing conditions.

Generally timbers can resist tensile stresses better than compressive stresses.

Both compressive strength and tensile strength are greatest along the grain.

Stress grading of timbers for structural use is now preferred to selection by species.

Pines and other 'softwood' species usually have lower stress gradings than eucalypt and other hardwood species. Refer to Standard Codes and local information.

Manufacture and availability

Both natural and man-made forests are now exploited for timber. Top quality timbers from natural forests are now very scarce, due to indiscriminate cutting in the past.

Many forests now produce fast-growing softwoods rather than the less productive hardwoods. Future uses and availability of timber depend on forestry policies and government controls, which have very far-reaching effects for the future. Locally grown exotic trees can have different qualities from the original product. Imported timbers usually arrive in balks and are cut to size locally. Local timbers are generally cut to scantling sizes in the original sawmill, usually located near the forest.

Long lengths and wide boards of best quality building timbers are now difficult to obtain and may need to be ordered well in advance of delivery dates.

Milling and marketing tends to concentrate on standard sizes of common components for domestic construction.

Australian eucalypt hardwoods are normally used while still partially green for framing purposes.

Selected and well-seasoned timbers are needed for joinery work. Quality furniture manufacturers stockpile timbers in order to ensure long seasoning and consistent quality for production runs. Some joinery timbers do not perform well if seasoning is accelerated by kiln drying.

This bulk store for salt beside the ocean spans 41 m and is constructed using glued laminated timber frames which are not affected by the salts corrosive to other structural materials.

TIMBERS

Water and atmospheric effects

Most timbers need to be kept protected from rain and excessive humidity to retain their original qualities.

Optimum moisture content for the climate and location in use needs to be a criterion for timber quality specification. Actual percentage of moisture content will vary with climatic and seasonal condition when in use.

Timbers shrink as moisture content decreases and swell as it increases. This is a continuing process. Dampness or inadequate ventilation of surrounding atmosphere can lead to decay caused by fungi.

Oils or paints can minimise the moisture absorption of timbers. Some species of timber are resistant to weather exposure and have long life without protection (e.g. American western red cedar and some Australian eucalypts).

Marine conditions can be resisted by some species and coastal and polluted atmospheres seldom affect timber's useful life. Most pines (softwoods) are liable to rot if left unprotected from weather. (Pressure-impregnated timbers are excepted.) Australian hardwoods tend to crack if exposed to weather, but retain considerable strength and have long life even if in the ground.

Steam can be used to soften and bend straight timbers to curved shapes for structural members furniture, boat building, etc.

Heat, temperature and other effects

Dry heat greatly reduces moisture content and causes shrinkage. This often leads to splitting and cracking near the ends of boards, or at nail holes.

The major shrinkage of timber occurs across the grain. Very little change in length of timbers occurs due to humidity or temperature variation.

Timber is a reasonably good insulator against heat, and because of its porosity and lightness has a low heat storage capacity.

Fire is a major hazard for timber, but some hardwoods and laminated members in structural sizes above 70 mm thick are slow burning and self-insulating by charcoal built up on the surface; they can be fire resistant after charring.

Timbers produce noxious fumes when burning which can create serious hazards in fire.

Ultra-violet radiation

Timbers generally perform best in their local habitat. High ultra-violet radiation in some locations can cause brittleness and cracking.

Cracked timbers are more vulnerable to fire risks than are solid timbers and allow moisture penetration.

Electrolytic and other special characteristics

Timbers generally are electrical insulators and do not react electrolytically with other materials.

Some timbers exude gum from veins after sawing. Other timbers cause staining from rainwater if left unprotected. There are numerous defects that can cause timbers to be rejected for structural reasons such as knots, shakes and splits. Ability to hold nails and/or screws varies greatly with species, and nailability near the ends is a feature of importance for floor boards, furniture, etc.

End grain does not provide a good hold for nails or screws. Hardwoods have much greater nail-holding power than softwoods.

Durability and susceptibility of timbers to insect attack are important structural considerations.

The insulation properties of timber contribute to the pleasing `feel' or tactile quality which make it very suitable for handrails, furniture, etc.

Acoustic properties

As normally used, timber is not an effective insulator against airborne or impact sound but is more absorbent than masonry and other hard materials. It can be used effectively in conjunction with other absorbent materials to minimise reflection of sound and help produce resonance in certain types of spaces.

See **timber.org.au** for more details.

External timberwork showing signs of deterioration at joints due to weather exposure and insufficient maintenance of paintwork.

Standards

AS 01. -1964	Glossary of terms used in timber standards
AS 02 - 1970	Nomenclature of Australian Timbers
AS 1080-1972	Methods of testing timber & moisture content
AS 1143-1973	Creosote for the preservation of timber
AS1148-1971	Nomenclature for timbers imported into Australia
AS 1262-1972	Code, for wood mosaic parquetry flooring
ASCA 31-1960	Code for laying parquetry flooring
ASCA 39-1963	Code for sanding interior wooden floors
AS 1495-1973	Preservative treated radiata pine cladding
AS 1540-1974	Timber frames & sashes for windows
AS 1604-1993	Preservative treated timber
AS 1606 & 1607-1974	Water repellant treatment of timber
AS 1684-1992	National Timber Framing Code
SAA.HB 44	A guide to the Timber Framing Code
AS 1694-1974	Code of Practice for Termite Protection
AS1720	Timber Structures Code
AS 2146-1978	Performance of timber window assemblies
AS 2147-1978	Code for timber windows in buildings
AS 2543-1.983	Nomenclature of Australian timbers
AS 3660	Protection of buildings from termites

CSIRO Publications

NSB's		
	BTF 04	Timber Fasteners
	34	Timber in domestic buildings
	42	Claddings in small buildings
	44	Traditional roof framing for tiled roofs
	45	Timber floors in dwellings
	103	Industrial timber floors
	121	& BTF 7 Timber fasteners
	165	Wall cladding
	168A	Overclad or paint

State Forests of NSW Publications

available from: 121 Oratava Avenue West Pennant Hills NSW 2125

TP-1	Forest products for domestic construction
TP-2	Sapwood & heartwood.
TP-3	Finishes for exterior timber surfaces
TP-6	Timber for external uses.
TP-10	Flooring timbers of NSW.
TP-11	Under Floor Ventilation.
TP-12	Floor finishing and maintenance
TP-13	Timber species properties & uses
TP-14	Termite control for existing buildings.
TP-19	Lyctus Susceptibility of timbers
TP-20	Kiln Drying.
TP-28	Density of Commercial Timbers.

Interior of a contemporary residence showing use of timbers for structure, flooring, wall linings, doors, chairs and tables.

Timber & Building Industry Association
PO Box 107 Surry Hills 2010
Leaflets TIL Nos :-
1. *Introducing Timber*
2. *Bathroom panelling*
3. *Timber framing*
4. *Decorative floors*
5. *Corrosion & timber*
6. *Wall panelling*
7. *Forest, wood, timber.*
8. *Stressgrading*
9. *Pool safety fences*
10. *Outdoor timber projects*
11. *Oregon & the outdoors*
12. *Timber, fire, structures.*
13. *Timber flame & fire hazards*
14. *Ordering timber*
15. *Preservative treated timber*
16. *Timber - Future supply- '93*
17. *Cypress*
18. *Hardwoods*
High & Dry- literature & builder's guide

National Association of Forest Industries
Publications available from www.nafi.com.au
Websites for above organisations should be referred to for up to date information and printed matter available.
Timber Manual
Timber Living Concepts Timber Frames
Pole House Construction
Protecting buildings from subterranean termites
Manuals 1. 2. 3 Multi residential Timber Framed construction
NSW Timber Framing Manual

ORGANICALLY BASED BUILDING BOARDS

Introduction
Much experimentation has gone on since 1900 in attempts to utilise some previously wasted timber and agricultural by-products. Many resulting products have proved very successful and are now widely used.

Most of these products overcome some of the unfortunately variable qualities of natural timbers. Most are marketed in the form of thin sheets and/or wide boards that can be cut and worked readily with wood-working tools.

PLYWOOD
Plywood is one of the oldest of timber products. Its antecedents go back to the Egyptian methods of making papyrus by layering reeds at right angles to produce a tough fibrous sheet.

Plywood is now a product requiring heavy powered machinery which can peel continuous thin sheets off a rotating log up to 3 metres in length. These sheets are then overlaid so the grain runs alternately at right angles in successive sheets.

The layers are bonded together with glue, under pressure. Finished sheets of plywood consist of three or more layers and can be from 3 mm to 25 mm thick, and some manufacturers make heavier sheets. Sheet dimensions are usually from 1800 x 900 mm up to 7500 x 2100 mm. For special uses giant sheets of 15 m x 2.7 m have been produced. The resulting sheets are dimensionally stable and resistant to cracking.

Plywood has remarkably good strength-to-weight ratio and is very resistant to shear stress. It can be used as a component of a structural member; as formwork for concrete; as flooring supported on timber framing; as cladding to buildings; decorative panelling; for furniture, etc.

Thin sheets can readily be bent to assume curved shapes. Surface veneers can be decorative timber grains of selected species if required and take stain or clear finishes.

Modern glues and surface coatings can produce plywods capable of resisting moisture, external weathering, wet concrete, and even marine conditions. Sheeting grades in plywood must be specified carefully according to the use intended.

Normal plywods for decorative and internal use will not be satisfactory if exposed to external weathering conditions.

The Plywood Association of Australia has a quality control identification stamp on sheets approved as follows: Exterior Plywood; Structural Plywood; Interior Plywood; Marine Plywood.

An indication of the toughness of plywood is given by its widespread use as concrete formwork. It is often reused many times and suffers frequent rough handling and transport as well as weather exposure in the process yet is readily cut and fitted using carpentry tools.

SOLID COREBOARDS
These products were popular from the 1930s prior to the widespread development of thick particleboards. They are still used for some types of solid core doors, especially fire doors.

They consist of a series of edge-glued timber strips, usually 25 mm to 50 mm thick, covered on both faces with plywood sheeting, glued on under pressure. The resulting panels are very stable and the finishing surface veneers can be decorative as well as structural.

Due to the relatively high density of these boards they are also used as doors for acoustic insulation, being far superior to hollow core or panelled doors in this regard.

FIBREBOARDS
High density smooth faced fibreboards were first marketed as 'Masonite' in the 1930's. and are still available as weather resistant external cladding material, but were gradually superseded by the thicker and more rigid lower density chipboards in various thicknesses during the late 20th Century.

Chipboards. A great variety of boards are now available, made from chips of timber fibres that are brought together with natural or synthetic glues under pressure to produce stable, even textured and smooth surfaced boards. Most domestic and commercial fit outs now incorporate large quantities of these boards. Many of the trees used come from plantation forests.

Large softwood posts and beams with waterproof plywood balustrade on a temporary footpath hoarding structure bolted together for ease of dismantling.

The process of manufacture uses pulped fibres to which are added glues and/or some waterproofing agents. A mat of the pulp is fed into a large press which has smooth press plates. Heat treatment is applied in ovens, followed by a saturated steam atmosphere for some hours to bring moisture content back to the desired low percentage.

Chipboards achieve their bond and strength from the multidirectional felting effects of the fibres. This produces a very even cross section of material with a slightly increased density at the surfaces where pressure has been greatest. Chipboard advantages are the availability in various sizes and thicknesses; with dimensional stability, smooth faces and ease of working, making them very useful for cabinet work using normal carpentry tools.

A disadvantage is their unsuitability for edge screwing. Without special treatment they are not water resistant and should not be used externally; close to steam or high moisture areas; or in commercial kitchens. Prolonged moisture exposure causes the boards to swell. The natural surfaces are not resistant to wear and tear, and for much cabinet work are finished with a sheet veneer or plastic laminate glued on.

Chipboards do not have the attractive grain, or strength in bending and shear which many timbers have, but their dimensional stability and large sheet sizes are great advantages for much cabinet work, for which special jointing and hinging devices have been developed. Screws with full length threads need to be used for fixings to chipboards.

Such boards are also used as flooring, for which special water resistant qualities are available. They are not fire resistant, and do not have very good shear resistance so heavy point loads should be avoided. A wearing surface such as vinyl or carpet is necessary for floors.

Medium Density Fibreboard (MDFB) is a recently developed smooth faced board available in a range of thicknesses from 3 to 32mm, It is particularly suited to working with mechanical tools to produce mouldings and curved shapes, being of even quality in all planes. MDFB is moisture resistant, but does swell slightly with humidity, hence needs expansion joints when used over long lengths.

BUILDING BOARDS

It is now superseding the coarser textured chipboards for good quality cabinet work, and replacing timber for much fit out joinery. All surfaces take paint equally well and do not need laminate protection. These boards are mostly made from quick growth renewable forest timbers, and consequently are environmentally acceptable

Laminated Veneered Lumber (LVL) also referred to as Engineered Timber, is a development intended for use in timber structures, in which large particles or veneers of timber are clearly visible. Various thicknesses and long lengths are available. They tend to be more structurally strong and stable than chipboards or softwood timbers, and with the aid of modern glues can be made up into efficient shapes such as for I beams. They are readily used for formwork, and have been adopted for structural and decorative uses.

Composite Boards have now been developed which contain three or more layers of different materials, not all of which are necessarily of timber origin. Their thickness is usually about 8 to 10 mm. The surface layers are often pre-finished with edges tongued and grooved for secure fixing. Some of these are used for flooring, and others as internal linings. The variety and range of uses seems likely to increase. These boards have been made possible through the development of high quality chemically based bonding agents, and may be one way in which valuable and scarce timber materials can be conserved. Local timber merchants or plywood suppliers should have information regarding varieties available in their area.

Softboards

The early softboards developed from the use of sugar cane fibres left after crushing at the mill. Now most softboards are made from softwood forest timbers. These boards have little structural strength or stability and need to be protected from moisture, hard knocks, etc. They are useful, however, for some acoustic applications, to absorb sound, and are available in several thicknesses, from 10 mm to 25 mm, and in various sheet sizes. Because of their low density and soft surface, pins can be pushed into them; and they are highly flammable.

Strawboards

Some compressed straw-type boards are available, approximately 50 mm thick. The straw can be glued together and enclosed in a heavy paper casing, giving a smooth surface finish, or it can be bound and glued to leave the natural straw visible.

These boards are structurally stable and reasonably light; they can be used for interior partitions and for ceilings. They have useful acoustic absorption properties and are resistant to fire. Details of these qualities need to be checked from test reports available from the suppliers.

Timber/cement Boards

By combining timber strips or shavings of very small cross section dimensions with cement slurry and compressing them together lightly, boards can be made which are dimensionally stable, fire resistant and have sound absorbent qualities due to cavities in the surface.

Unfortunately resulting products have a drab grey cement colour, and this limits their appearance value. Spray painting is not very successful and reduces their acoustic properties. However, these boards have been used widely in suspended and other ceilings and provide an interesting textured effect.

Timber Products Standards

AS 1684.1-2-3-4:1999	Residential timer framed construction
AS 1720.1-2-4:1997	Timber structures; Properties; Design; Fire.
AS /NZS 1859	Reconstituted wood based panels
AS 1860-1991	Particle board flooring
AS /NZS2269-1994	Plywood Structural
AS 2270-1979	Plywood & blockboard for interior use
AS 2271-1979	Plywood & blockboard for exterior use
AS 2272-1979	Marine plywood
AS 2329-1980	Mastic adhesives for fixing wallboards
AS 2458-1982	Hardboard
AS 2459-1982	Organic fibre insulating board
AS 2754	Adhesives for timber products
AS 3600.1:2000	Termite management; new buildings.

Plywood Association of Australia Publications

3 Dunlop St. Newstead Qld. 4006
Other major plywood manufacturers also have literature available.
Structural plywood wall bracing design manual
T & G structural plywood for residential flooring
Low profile floor systems
Timber tops for concrete slabs
Structural plywood for commercial & industrial flooring
Plywood in concrete formwork
Design guide for plywood webbed beams
Structural plywood -design manual
Joint Aust & NZ standard for structural plywood
Plywood - Ideal for professional or handyman
Plywood- the only engineered wood panel
Facts about plywood
Plywood webbed structural beams for domestic housing

Some of the textures available in timber-cement boards which have useful acoustic and decorative qualities. These boards no longer locally available.

Engineered timbers can now be made into efficient I shaped beam sections.

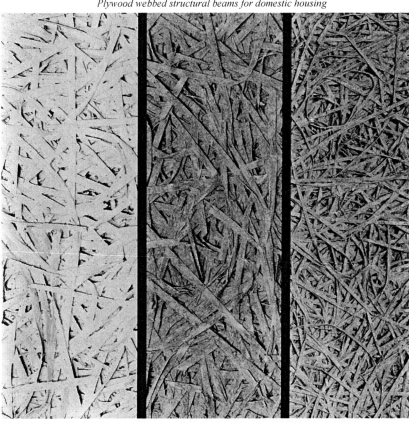

BRICKS AND BRICKWORK
History; raws materials; manufacture; quality and types; dimensions; growth; specials; properties; fire resistance; references.

INTRODUCTION

Bricks, made by shaping a plastic mass of clay and water which is then hardened by drying and firing, are among the oldest and most enduring of mankind's manufactures. Until comparatively recent times the clay was dug, the bricks were made and kiln set or drawn by human labour with only minor help from animal power. About 100 years ago the first effective machines for brick production appeared and the trend towards mechanisation of clay winning, making and handling operations has continued at an increasing pace to the present day.

As the first European settlers arrived in Australia in 1788, with scarcely any building materials or tools suited to their needs, they quickly turned to the manufacture of bricks once suitable clay was located close to the Sydney Cove settlement.

Since then Sydney has become largely a brick city and the manufacture of bricks has extended to many cities and towns. Isolated homesteads were often constructed of bricks manufactured from local clays fired on the site by itinerant brickmakers whose expertise has now been rendered unnecessary by modern transport.

As explorations led to the establishment of new towns, and ships from Britain found new ports - settlements that later became the capitals of States were founded, and these too began to use bricks as suitable clays were discovered locally.

The Australian pattern parallels much earlier developments in the use of bricks. Areas where timber and stone have been scarce or unsuitable have developed brick technology. The earliest examples were sun-dried bricks as still used in many countries today.

Egyptian builders considered brick to be inferior to stone and used it only in unseen locations.

Early Persian uses of brick are well known and some very ancient examples still exist as testimony to the durability of brickwork.

The Chinese had well-fired bricks in use in the third century BC when the Great Wall was constructed.

The Romans used brickwork for structure and decorative purposes. The circular and segmental arch form, often of brickwork, became an important element of their architecture wherever they conquered. They spread the art of brickmaking into Europe and England.

The thin, red, Roman-style bricks are still a feature of many Italian cities today.

The shape and size of bricks varies according to local history and traditions. Once these are firmly established, it is useful to retain standard sizes in an area so that alterations can be made to existing brickwork with a minimum of trouble. In this way brickwork has proved to be one of the most flexible of building materials, which often displays strong local characteristics of size, texture and colour.

Mortar for Brickwork

The performance of brickwork over a long period depends very much on the properties of the hardened mortar, and the quality of workmanship of the bricklayers. The mortar therefore needs to be both plastic enough to use and suitable for its ultimate purpose.

The lime content provides resiliency and self-healing qualities allowing some adjustment to minor movements. The cement imparts hardness, adhesive qualities and compressive strength. The sand influences workability, shrinkage and ultimate strength. Admixtures are frequently used to modify colour, workability or setting properties but must not be used to replace the lime content.

The following mortar selection table is taken from BDRI Technical Note - Mortar for Bricklaying.

Application	Cement	Hydrated Lime or Lime Putty	Sand
High durability where in contact with the earth - foundations, retaining walls etc.	1	1/10	3
Maximum bond - Structural brickwork	1	1/4 to 1/2	4 to 41/2
General Purpose - Severe exposure	1	1	6
Low compressive stress - Not suitable for freezing	1	2	9
Single storey interior or protected exterior use	1	3	12
Single storey, interior or protected exterior	-	1	3

NOTE: In the above table the material proportions are measured by volume, preferably using a gauge box for consistency.

Convict made and laid soft sandstock bricks showing units in arches hand rubbed to shape.

Late 19th-century decorative masonry walling incorporating contrasting coloured and moulded bricks, sandstone and terra cotta panels.

RAW MATERIALS FOR BRICKS

Conventional bricks can be made of a wide range of clays and shales. These have one thing in common: they are always composed of secondary or water-containing minerals produced by the action of weathering agents (water and air) on primary or igneous rock minerals such as feldspars and micas. These secondary clay minerals are compounds of alumina, silica and water, with minor amounts of lime, magnesia, soda or potash. Iron compounds, usually the oxides, hydroxides or carbonates, are nearly always present as impurities in brick clays, and account for most of the wide range of colours found in finished bricks.

Clays containing up to 3 per cent of iron oxide give white to cream or buff colours, which change to pinks and reds as the iron oxide content rises to 8-10 per cent. By adding manganese dioxide in amounts of from 1-4 per cent, a range of grey and brown colours can be produced.

More important than their chemical composition is the fact that, when mixed with water, the clay minerals give a plastic mass which can be shaped by pressure to form a brick. Unlike a mixture of sand and water which also can be moulded to a desired shape, the clay-water mass retains its shape as it dries, and when fully dry is usually quite strong.

In Australia most building bricks are made of shales or shaley clays. In Brisbane, for instance, these are drawn from the clays or claystones associated with local coal measures; and in Sydney pressed bricks are made of lightly weathered shales occurring west and south of the city. Similar hard shales are used in Hobart. In Melbourne, Adelaide and Perth a softer, more weathered type of shale predominates.

In all parts of Australia recent years have seen a rapid increase in the production of extruded bricks, which generally require softer material, usually obtained from transported or secondary clay deposits found near the surface. This has caused old brickworks to move from shale deposits, and new plants to be built convenient to suitable clays and transport services.

MANUFACTURING PROCESS

Generally there are three stages in manufacture:
(1)   winning the clay and preparing it
(2)   shaping it
(3)   drying and firing process

(1) Winning the clay

Raw materials for commercial brick manufacture are now almost wholly won by mechanical means from pits located on suitable deposits of clay and/or shale etc. The excavated materials are crushed and milled and screened for particle size. The pits and kilns traditionally were adjacent, but now the clays are often transported and blended to maintain consistency of the product.

(2) Shaping the product

Four methods of shaping are commonly used:

(a) 'Wet Process' Hand-moulding (Sandstocks) - sand-dusted wooden moulds.

A very wet or plastic clay needed - dried for 2 or 3 days and burned in a kiln.

Type of brick likely to vary in size and be slightly twisted and warped due to excessive moisture content and drying out. Surface texture pleasant.

(b) 'Wet Process' Extruding - a continuous column of clay dried and burned as for Sandstocks

- a stiffer clay used

- no frogs - poor arises - textured surface on two faces.

Dry pressed bricks - modern method

(c) 'Dry Process' - applying a high pressure to a ground and dry type of clay in a mechanically operated press which presses the bricks in steel-lined box - i.e. Dry Process.

(d) 'Wet Process' Machine Moulding - uses a fully plastic clay similar to that used for hand-made process, the difference being that clay is mechanically 'squeezed' into a wooden mould and then dried and burnt in the same way as for hand-made bricks.

3. Drying and Firing

Water contained in the clay must be dried out before firing.

BRICKS

The drying treatment depends on the method used to make the bricks and on the kilns used. Extruded bricks have a higher water content and denser structure than pressed bricks. The holes usually incorporated reduce the thickness of clay through which water has to escape, thus speeding production and reducing risks of cracking.

Three methods of burning are:

(a) **Clamp** - Mainly confined to South-East England - subject to excessive variation and distortion - not used in Australia.

(b) **Intermittent Kiln** - Capable of a very close degree of control and high temperatures.

- expensive method of burning

- generally used where bricks of special quality or colour are wanted.

(c) **Continuous Kiln** - Most economical method of production - greater proportion of bricks in this country are made in this way.

Delivery

Methods of handling bricks have evolved rapidly in recent years; and although a large percentage of all bricks is still loaded on and off flat-top motor trucks by hand, several systems of mechanical handling are being used.

In some pack systems the required number of bricks is stacked to give a block of the correct shape and number and the pack is then picked up by a special hydraulically-operated truck hoist. On the building site the system works in reverse, to deposit the load of bricks on the ground.

Each time a brick is handled it is likely to be damaged by losing its sharp edges or by cracking. Care with timing and locating site deliveries, minimum handling and the use of proper barrows, etc. are all important in producing good-quality brickwork.

Exterior view of intermittent brick kiln sealed for firing.

View of kiln interior with bricks open stacked to help uniform heat penetration

QUALITY OF BRICKS

High compressive strength bricks are -

(a)   Of even texture and colour characteristics.

(b)   Have a clear ringing sound upon impact.

(c)   Have sharp unbroken and square arises.

(d)   Able to resist a pressure of at least 20 MPa.

In addition to the manufacturing terms that are used to describe bricks there are some terms used which refer generally to the quality of the bricks as they come from the kilns, although with many new control techniques there is very little variation in quality from any one plant or batch.

These terms are as follows:

Callows - Unburnt, usually pale straw to whitish colour, structurally weak, poor in quality, deteriorate quickly. (Do not confuse with pale-coloured face bricks.)

Clinkers - Overburnt - may have fused or run together under extreme temperatures in the kiln; very hard - usually misshapen and of irregular size.

Commons - Ordinary bricks usually made by dry process and are the average product from the satisfactory burning of a continuous kiln; broken and chipped arises, shrinkage cracks etc. included.

Selected Commons - Common bricks of better than average quality, with clean, sharp arises; may be used for face work. O.K. Commons - A further division between Commons and Selected Commons.

O.K. Face Bricks - Selected bricks of even colour and texture due to maximum controlled burning and treatment; should have a clear ringing noise when hit together.

Sandstocks - Bricks shaped by placing the clay in a sand-lined mould. The resulting bricks do not have sharp, clear arises and corners.

Fire Bricks - Made of fire clay and are used to withstand extreme temperatures, furnaces, etc., fire box linings.

Extruded Wire Cut Bricks - Term refers to the process of manufacture.

Glazed Bricks - Bricks have a hard, glass-like transparent coating produced by the throwing of salt into the kiln during the burning, or coating one face with a glazing compound before firing.

Moulded or Special - Bricks of special size or shape for decorative or other special purpose.

Colour of Bricks - Dependent on impurities in clay such as lime and metallic oxides, and by temperature of the burning process. These can vary greatly without being detrimental to quality.

The variety of colour and texture, and the individual variations between bricks of each batch, are attractive visual qualities.

FROGS IN BRICKS

The depression in one bedding face of pressed bricks is usually called a 'frog'. Its purpose is to make bedding down into the mortar easy when laying, and to increase resistance to shear slip along the bed joint.

HOLES IN BRICKS

In extruded bricks holes usually replace frogs, as this suits the manufacturing process and helps the kiln heat to fire the clay evenly throughout.

BRICK DIMENSIONS

The prime factor controlling all brick dimensions is the size and weight of a unit that can be grasped in one hand. The secondary factor is the modular relationship of length to width - usually 2:1, including allowance and/or joint thickness. With these factors, a great variety of sizes is possible.

The Romans used a very flat brick approximately 50 mm thick. Similar bricks are still commonly used in Italy. Australia's early brickmakers followed English 18th-century practice and made bricks sized approximately 9 in x 4 5/16 in x 3 in.

With metrication, in the Twentieth century, rationalisation of brick dimensions caused English sizes to reduce and Australian practice to hold closely to its traditional brick size. A new metric modular brick was introduced which is longer and narrower than traditionally-sized bricks, changing the 2:1 ratio to 3:1, but this dimension is not so readily suited to clay brick manufacture and is more commonly made in cement products.

The nominal sizes for Australian clay bricks now stand at: Metric Standard, 230 x 110 x 76 mm.

Metric Modular, 290 x 90 x 90 mm.

BRICKS

Contrasting sizes and qualities in dry pressed bricks in current manufacture.

Extruded bricks from differing factories showing variations in colour, texture and dimensions.

Standard and metric size extruded bricks with various perforation patterns which help produce even baking.

BRICK EXPANSION OR GROWTH

20th Century research in brickwork has shown that local extruded bricks expand in three dimensions after removal from the kiln. Laboratory and field measurements indicate a high early rate of expansion during the first month is common, followed by an almost constant growth of approximately 0.006 mm per annum for five years or more.

This important Australian research has been published by the CSIRO Division of Building Research and the Brick Development Research Institute.

Results of this expansion have been clearly evidenced in many buildings and designers now have to allow for this with control joints. Growth can approximate 20 mm in a straight wall 30 metres in length.

SPECIAL BRICKS

In addition to the normal rectangular bricks for building walls many other shaped bricks are produced for special purposes, such as shaped sill bricks, decorative moulded bricks and paving bricks.

Paving bricks especially need to be suited to the way they are to be used, as some faces are much softer than the intended wearing face. Wrong choice or placement can lead to rapid deterioration of the pavement. Paving bricks are particularly suited to exterior areas where a non-slip surface, pattern and colour are desired.

Face textures are applied to some bricks and this technique has been used extensively with extruded bricks.

PROPERTIES

Technically, the most important properties of bricks are their strength, their absorption properties, and their insulation against sound and fire.

Strength

The compressive strength of building bricks in Australia varies from about 20 MPa to 80 MPa. In many applications (e.g. in brick veneers, for infill panels in frame structures, and for loadbearing walls in small buildings of one or two storeys) this strength is not required, but it can be useful in engineering applications.

Generally speaking, extruded bricks, even when heavily perforated, are stronger in compression than solid pressed bricks. By reinforcing brickwork with steel rods and wire, tensile strength can be developed in brick walls and this is now being exploited by factory fabrication and road transport of large brick panels for cladding buildings in the USA. Handling is similar to pre-cast concrete panels.

Brickwork cracking vertically near corner due to brick growth in long straight return wall.

BRICKS

Absorption Properties

The power of the brick to soak up water is one of its most useful characteristics. When the bricks are being laid their suction plays an important part in developing good bond with mortar, and in the finished building the absorption of rain by the bricks reduces run-off which may otherwise cause trouble at flashings and around windows or other openings. The water absorbed during wet spells is harmlessly evaporated again when the weather clears.

In freezing conditions up to 7 % absorption is considered to be safe for brickwork.

The absorption properties of bricks affect the mortar in which they are laid so local practice varies regarding wetting of bricks to achieve optimum results in a built wall.

Insulation

Sound. Insulation against air-borne sound depends more on the mass or weight of the wall than on anything else. A solid brick wall sets the standard for sound-proofing in both residential and commercial construction.

Fire. The fire resistance of brickwork is high, because the materials are inherently resistant to fire. Its low thermal conductivity ensures that heat is not quickly transferred through brick walls. Fire rarely damages brickwork in a building, although thermal expansion of unprotected steel columns, girders or roof trusses may cause displacement or even collapse of brick walls. See `Comparative Tabulations' for Fire-Resistance Ratings for brick walls - Page 109.

NON-CLAY BRICKS

Twentieth-century technology has introduced bricks and blocks made from materials other than the traditional shales and clays and these have been widely adopted in many localities.

The principal ones are cement and concrete blocks of which there are numerous types, shapes and sizes. Because of their kinship with cement and concrete products they are dealt with generally under those headings.

Also popular in some areas are calcium silicate or sand-lime bricks which are usually whitish in colour, smooth finished and similar in texture to cement bricks. Their characteristics, however, vary considerably from clay/shale bricks and do not have comparable colour range or proven durability. Extreme care should be taken in using silica bricks in exposed locations and case studies made to check performance. Use of silica bricks for the internal skin of cavity walls where clay bricks are used for external skin is not recommended as this combination has demonstrated some unusual cracking patterns in apartment-type buildings, and more investigation of this phenomenon is needed.

CONCLUSION

Substantial increases in brick production in recent years are ample proof of the rising demand for the products of this very old industry.

Brick manufacturing is being recognised as a large-scale chemical engineering operation. Modernisation of plant and the application of up-to-date methods of materials handling have increased efficiency and improved both the range and quality of products available.

Methods of using bricks are also undergoing modernisation, as part of a general reappraisal of the strength and performance of building materials. Massive brickwork has been recognised as a wasteful application of a potentially strong material.

Extensive experimental investigation of brickwork strength has resulted in simple engineering design procedures which lead to economical structural applications for thin brick walls. Used in this modern way, brick construction continues to be structurally useful and economically competitive, whilst retaining its traditional virtues.

Use of steel reinforcing rods or wire in bed joints can greatly enhance the tensile strength of brickwork and further developments are under way with factory fabrication of large reinforced brick panels similar to pre-cast concrete panels.

Despite the collapse of steelwork in fire, the brickwork here was little damaged and protected the steel columns.

Specially shaped and textured bricks for paving are suitable for both wheeled and foot traffic.

BRICKS

SUMMARY OF BRICK CHARACTERISTICS

1. Structural Strength
Bricks can satisfactorily resist compressive and low shear stresses, but they are not able to resist tensile stress. The compressive resistance characteristics of walls are dependent on quality of mortar used, the skill of the layer, and the bricks themselves.

2. Manufacture
Modern manufacture can produce good quality bricks equal to a given sample for almost any number required.

3. Water and Freeze Thaw effects
Bricks absorb water and consequently water-borne chemicals. Well-burnt bricks usually resist normal atmospheric chemical attack. The freeze/thaw effect on wet bricks is the most serious cause of physical damage to brickwork. Moisture often penetrates porous bricks by capillary effects. In severe marine environments bricks should be well burnt and dense to avoid destructive salt attack.

4. Temperature Effects
Heat and temperature variations above the freezing level of water have little effect except for expansion and contraction in long straight walls. Heat is stored in brick walls and slowly radiated when the atmospheric temperature drops.

5. Ultra-Violet Radiation
Ultra-violet radiation has no effect on the colour or life of bricks.

6. Electrolytic and Other Special Effects
Bricks do not produce any positive or negative electrolytic reaction.

A disadvantage with modern bricks is their tendency to grow in length after they have been built into a wall. This can produce serious problems in many situations unless designed for initially.

7. Acoustic Qualities
Due to the dead weight of brickwork, the acoustic insulation of a properly constructed wall is adequate for most normal domestic and office requirements.

8. Reinforced Brickwork
Brickwork may be reinforced by laying galvinised wire mesh in the bed joints. By this means steel lintels can be eliminated over many door and window heads.

A recent development to repair or strengthen masonry walls, inserts patented helical metal ties grouted into pre-drilled holes.
See www.helifix.com.au

References
The Glossary of Building Terms (SAA HB 50) or its predecessor, a Glossary of Building and Planning Terms should be referred to for Australian bricks and brickwork diagrams, and related technology.

Standards
AS 1225-1984	Clay building bricks
AS 1617-1993	Refractory bricks and shapes
AS 1653-1985	Calcium silicate building bricks
AS 2904-1986	Damp proof courses and flashings
AS 2975-1987	Accessories for masonry construction
AS 3700-1988	SAA Masonry Code (with supplements 1 to 5)
AS 4678:2002	Earth retaining Structures
CBPI TN 21A-B:198	Clay brick walls & fences
SAA HB33-1992	Domestic open fireplaces

CSIRO NSB's & Building Technology Files
62	Brick veneer construction
126	Clay brick manufacture
134	Movement in brick buildings
135	Design of expansion gaps
119	Brick testing
BTF 05	Cleaning brickwork

The Clay Brick & Paver Institute (see websites for addresses) has a catalogue of publications with prices, including the following valuable papers :
Design Manual 2 The basics of brickwork part A-Materials
Design Manual 5 Fire resistance levels for clay brick walls
Paver Note 1 Specifying & laying clay pavers
Tech Note 2A Glossary of brickwork terms
Tech Note 8C, Brick houses in bushfire prone areas
Tech Note 10A Dampness in brickwork
RP 6 Frost resistance of brickwork in the Australian Alps
RP 9 Moisture expansion of clay bricks
Guide to Masonry Construction.

Severe brick erosion due to salt in a marine atmosphere. Earlier attempts have been made to patch erosion.

LIME, CEMENT, MORTARS, PLASTERS AND RENDERS

Rock, hydrated, hydraulic limes; cement; sands; lime mortars; composition mortar,
bush sands; plasters, rendering, cement render; additives, gypsum; fibrous plaster,
plaster boards, fibre-reinforced cement.

INTRODUCTION

In the history of building technology, setting agents have long been mixed with sands and water to produce a bedding of mortar for masonry work, and plaster for coating walls.

The setting agents traditionally used have included some form of lime or gypsum. More recently Portland cement has been used as well.

The long history of lime in relation to masonry walls shows that it is a very important basic resource for the industry. Its characteristics need to be understood and distinguished from those of other binders such as gypsum and cement.

LIME

Production

Lime is produced from limestone, chalk, coral, sea shells and other calcium-rich resources by heating them to a high temperature in a current of air in a special type of kiln so that carbon dioxide is lost and the result is calcium oxide or quicklime.

This process has been exploited by man as long as recorded history. In Australia the collection of oyster shells and the production of lime for building purposes occurred within the first few months of white settlement.

The raw material varies from place to place, both in its chemical composition and physical properties. The quicklime produced in different locations will vary somewhat in its behaviour, its speed of slaking, for instance, or its yield of putty when hydrated. If clay is present in the raw material, compounds are formed during burning which have setting properties similar to those in portland cement.

Rock Lime

Until recent years rock lime was hydrated on the building site to produce lime-putty; but the danger to humans from the corrosive effects and heat produced, are serious hazards in this process.

Hydrated Lime

Lime is now usually processed in factories into 'hydrated lime', which has had the correct quantity of water added to slake or hydrate the rock lime. By this means, a fine white powder is produced which is marketed in large paper bags. This product can be used with water for mixing into mortars or plasters to produce a material that is plastic and sets slowly as it dries out in the atmosphere.

Bagged lime must be kept dry before use.

Hydraulic Lime

Hydraulic lime is a type of cementitious lime which will set and harden under water in a manner similar to that of Portland cement. These limes are used widely in Europe for mortars in masonry, especially in underground or wet locations, but they are not popular in Australia.

CEMENT

In present-day building terminology, cement usually refers to the product Portland cement, which is the name given to the modern factory-produced material. It is pale grey in colour and derives its name from its similarity in appearance to English Portland stone when first marketed in the early 19th century.

Cement is now marketed in 40 kg bags and manufactured to strict controls in accordance with appropriate national standards. Portland cement is manufactured by burning a correctly proportioned mixture of limestone and clay (or of minerals of a similar chemical type) in a rotary kiln. At the high temperature of the kiln chemical reaction between the constituents occurs so that the resulting product is a mixture mainly of silicates and aluminates of lime.

The product from the kiln is in the form of semi-fused nodules known as Portland cement clinker. This clinker is ground finely, together with a small addition of gypsum, into the powder known as portland cement.

The gypsum is added to retard the set; without it the cement would be flash setting, but the quantity added is carefully regulated so that there is no risk of the expansion reaction peculiar to gypsum producing an unsound product.

Portland cements in Australia are made to comply with the Australian Standard for portland cement. This specification includes two types known as:
- Type A, normal portland cement, and
- Type B, rapid hardening cement.

As the name implies, the latter has to meet the requirements of a more rapid development of strength; however, the setting time is not necessarily faster.

The normal cement is grey in colour, but white cement can be obtained at a greater cost. It is not desirable to try to colour a cement on the site by mixing with a pigment, as it is not possible to be sure that an even distribution of pigment is obtained. Any colouring should be carefully measured and factory controlled.

The hardening of a mixture of cement and water is caused by chemical reactions between the two, resulting in the liberation of a certain amount of free hydrated lime and the formation of a colloidal complex. The latter can be compared with a flux, cementing itself together or acting as a cement between sand or other aggregate mixed with it.

Trucks loading bulk and bagged cement at factory. Note also the corrugated asbestos cement cladding to walling.

A characteristic feature of a binder of this type is that it shrinks on drying. It is not possible to prevent this phenomenon, but its effects can be minimised in so far as mortars and plasters are concerned, by extending the cement with sand using a mixture of cement and lime instead of cement alone, and allowing time for drying out between the applications of successive plastering coats.

The rate of hardening of cement is affected by the temperature. Little or no hardening will take place if plaster is frozen while still plastic. It should therefore be protected from frost. At the other extreme, it is advisable to prevent an excessive removal of water by evaporation in windy or hot weather, or removal of water by suction on a high-suction background.

The main advantages of portland cement are:

(a) it is a standard material readily available, of relatively constant properties;

(b) it develops considerable strength at early stages after mixing with water;

(c) because of its packaging and marketing it is handled easily for building purposes.

MORTARS

The disadvantages of Portland cement are:
- **(a)** its shrinkage movement on drying out after hardening;
- **(b)** the drab grey colour it tends to impart to most cement/sand mixes;
- **(c)** its limited bag life, especially in humid or moist conditions.

SANDS

Mortars, plasters, renders and cement products all incorporate sand as an ingredient. As sands available vary with locations, obviously the quality and content of the sands for building purposes will vary. They must all, however, be free of vegetable matter.

Several types of sand are usually produced by the bulk sand suppliers to suit differing needs in the industry. Coarse, sharp sand is frequently specified and used for concrete work, while 'plasterers' sand is usually finer and therefore more readily applied by hand methods. Sand for mortars may contain small quantities of clay and be coarser than for plastering. The preference of some bricklayers for 'bush' sand rather than washed sand reflects the improved workability the clay content can produce in mortars. However, clay smears or-streaks on brickwork can be difficult to remove. Coarseness and sharpness of the sand granules influence the workability and surface finish, particularly for cement render.

Local practice needs to be checked and samples tested where particular qualities or high compressive strength are required. In all cases sea sand or sands containing salts should be avoided. However, sands mined from semi-tidal lakes and rivers or seaside sand dunes are often used after washing, with satisfactory results.

Standard codes endeavour to define sand quality by proportions of sand grains passing certain sizes of sieves, but for some local situations, this method has proved to be unrealistic.

MORTARS

Mortar, in building, is a mixture of inert siliceous material (sand) with a binder and water brought to a plastic state which on drying out hardens into a stone-like material.

Mortar is used for the bedding and jointing of masonry (brickwork, stonework, concrete blockwork, etc.). The solids are mixed with just sufficient water to produce the necessary chemical reaction and plasticity for working. This can be done in small mechanical cement mixers, or by hand with a 'larry' or hoe-type tool.

Lime mortars are slow hardening and do not have a strong adhesion to masonry. They are usually supplied from tip trucks ready mixed in bulk plastic form and can be held on site for long periods if protected from the weather.

Lime mortar usually consists of one part hydrated lime or lime putty to two or three parts of sand by volume.

Such mixes are easy for bricklayers and masons to work with but limit the height of work that can be erected in one day due to the slow hardening characteristics. Drying is promoted largely by absorption of moisture into the masonry.

For stonework carrying heavy loads, practice was to keep the stones uniformly square and lay them with fine joints about 3 mm thick. This work is known as ashlar masonry; it minimises the tendency of mortar to squeeze out of thicker joints such as are normally used with brickwork.

Cement mortars are quick hardening and have a strong adhesion to masonry, together with a high degree of shrinkage occurring during drying. Setting occurs quickly after wetting; so mortar containing cement should be used within two hours, unless setting retarders are incorporated.

Cement mortars usually consist of a 1:6 cement/sand mix and sometimes are 1:3 and 1:4 in underground waterproofing situations. The cement-rich mixes become very hard to work and create excess shrinkage on drying, thus leading to cracking in the brickwork and/or the end joints.

Because of the quick hardening properties of cement mortar, masonry can be laid up continuously in most normal situations without squeezing out the lower joints.

The straight cement mortars tend to be a drab grey colour, which often spoils the effect of masonry colours. This unwanted drabness has led to the use of coloured composition mortars for exposed brickwork.

The CSIRO Building Construction & Engineering Division recommends a 1:3- cement/sand mix for waterproofing in underground locations and 1:6 general proportion of cement mortars.

Mortars which are rich in cement tend to cause unsightly cracking of masonry units due to the unnecessary tensile strength of the joints.

Composition Mortar (Compo)

As a compromise, to incorporate the advantages of the lime and cement mortars together, a mix known as 'compo' is often used. For normal conditions, portland cement is used as an addition to a lime/sand mix. The presence of lime in such a mix gives good working qualities and reduces both the rate of hardening and the final strength. Although the mix still shrinks on drying, the slower hardening rate tends towards a less harmful type of cracking.

The normal method of making this type of mix is to add the appropriate amount of portland cement just before use. The mix for brickwork is 1:1:6 - cement:lime:sand (by volume) - and prepared in the following manner:

To prepare a 1:1:6 mix, coarse stuff should be prepared by mixing 1 vol. of lime with 6 vols. of sand, and just before use, 1 vol. of cement should be added to 6 vols. of lime mortar.

Again, these mixes need to be used soon after mixing. Coloured mortars are also available, as for lime mortar, suitable pigments being added at the mixing plant. A wide range of pigments is available, and selection of an appropriate coloured mortar is especially desirable with brickwork exposed as face work.

Compo mortars can be made of these as before, but the cement content may deaden the colour a little.

Bush Sand Mortar

A common practice in some localities such as Sydney is to use cement and 'bush sand' in a 1:5 mix. The bush sand contains clayey loam that gives the desired ease of working for mortar, but it also produces a shrinkage factor. For domestic work this is not as serious a defect as it is in large-scale buildings.

Recommended Mortars: See item under Brickwork or refer to BDRI Technical Note - Mortar for Bricklaying.

Sand heap and equipment for mortar and plaster mixing on a building site.

PLASTERS

Plaster can include mixes of lime/sand (lime plaster) or lime and gypsum, or combinations of these. Importantly, plastering is an internal operation, finished invariably with a smooth texture.

The old type of plastering was carried out almost entirely in lime mixed with sand, and it frequently incorporated a liberal amount of hair. A feature of this work was that plenty of time was permitted between the application of successive coats, which were left undisturbed during the drying and hardening period.

The conditions under which a high proportion of present-day plastering has to be done permit neither of these factors to operate. The work is expected to be completed quickly, and it is often subjected to mechanical disturbance during the critical initial stages from other building trades working on the same building, and also from traffic and other vibrations.

By the introduction of new materials, or by the appropriate combination of new and old materials, plastering mixes can be chosen to satisfy the most severe modern conditions, but it is only by a proper appreciation of their potentialities and limitations that the best results are achieved and defective work is avoided.

There are three main groups of plastering materials, namely those based on lime, on portland cement, and on gypsum and anhydrate plasters.

The materials in the three main groups can be used alone or in certain combinations. Lime is used in admixture with portland cement or with one of the gypsum plasters. Plaster of paris and retarded hemihydrate gypsum plaster may be used with lime; anhydrous gypsum plasters are designed specifically for use with lime and sand for backing coats, although they can be used alone.

Keene's cement and similar hard plasters are designed primarily as finishing plasters, to give a particularly smooth, hard finish and on that account are not normally mixed with lime. Anhydrate plasters are incompatible with lime, which should in no circumstances be added to them.

Portland cement should never be used in admixture with a gypsum or anhydrate plaster because these materials react in the presence of moisture, the reaction producing a gross expansion.

All of the permissible variations can be used for interior work, but gypsum and anhydrate products are not suitable for use in exterior work, as gypsum is soluble to an appreciable extent in water; therefore, unless the material is given protection from the effects of water weakening, erosion and finally disintegration will occur.

Apart from these restrictions the choice of a particular plastering mix will depend on the nature and hardness of the surface desired, the time available for the work, the cost, and the nature of the background on which the plaster is to be applied. Certain types of background - e.g. gypsum plaster board - may be finished satisfactorily with only restricted types of plasters.

Lime Plasters

Plastering mixes based on lime having little or no hydraulic properties are not widely used today. They have no set, and harden by drying and the carbonation of the lime, brought about by the carbon dioxide in the atmosphere. This process is slow and uncertain and penetrates the body of the material only after a very long time.

Plastering with lime has been done, and still can be done, satisfactorily by a few tradesmen, but it cannot be hurried, and on this account is not suited to present-day requirements. Further, it is sensitive to shock or vibration during the early stages after application and may easily be affected adversely during this period, although it may not fail in a way that is obvious until much later. Its uses should be restricted to restorations.

Cement Plasters

Mixes based on lime but incorporating a cement addition overcome the difficulties already mentioned whilst retaining the workability of mixes containing lime.

Both lime and cement shrink on drying and undercoats containing cement, lime and sand have appreciable drying shrinkage. For this reason each coat must be allowed to dry thoroughly and so substantially complete its drying shrinkage before the following coat is applied.

Mixes of cement and sand with little or no addition of lime are used for undercoats for some gypsum plasters.

Cement and sand mixes are harsh to apply, particularly when a clean sand is used, and are in addition unnecessarily strong for most purposes. Their drying shrinkage, combined with their high adhesive strength, causes stresses to be set up both in the plaster and in the background to which it is applied which may cause defects in a background material weak in tensile strength. Their use should be restricted in the main to positions exposed to dampness or where maximum hardness and resistance to knocks is required.

RENDERING

Rendering is a term used for the act of applying plaster or cements to the naked surface of masonry walls. Present-day practice frequently eliminates the use of lime in plastering, with some serious results. Because of good adhesion and quick drying properties, sand/cement mixes can be applied in one coat thickly enough to produce a smooth surface over most masonry walls. This process is now usually referred to as rendering the wall.

Cement Render

For cement render the mix usually consists of 1 part cement to 4 parts sand, by volume, which produces a harsh material to work. The resultant cement-based coating, about 10 to 12 mm thick, shrinks as it dries out and usually crazing or fine cracks develop all over the surface.

If the base concrete or masonry is rough or absorbent enough, the render will hold to it, but if it is non-absorbent or smooth the shrinking stresses may cause patches of render to lose adhesion and become loose or 'drummy'.

This treatment of walls produces a very hard surface that can take rough usage, but it is not suitable for high-quality finishes or mouldings such as are used in traditional work.

The surface texture depends on the type of trowel used in the finishing process, a steel trowel producing a smooth surface and a wood float giving a rougher surface.

By using lime fatteners the harshness and crazing of the strong cement mixes can be reduced and ease of application increased, without seriously reducing the wearing surface qualities.

ADDITIVES OR ADMIXTURES

Proprietary additives based on petroleum and other chemicals have been developed to 'plasticise' or retard many cement-based mortars and plasters, but they are readily misused and often misunderstood.

As the quality, finish and life of mortars and plasters depend very largely on the proportions of the constituent materials, these additives should be used very cautiously. They have not survived the long-term testing needed for permanence in building terms, and their ultimate effects have yet to be demonstrated.

One of many textures possible using cement renders externally.

GYPSUM OR PLASTER OF PARIS

Gypsum is a naturally occurring mineral which is quarried or mined. Its composition is calcium sulphate with two molecules of combined water, and its chemical formula is $CaSO_4\ 2H_2O$. When it is heated to a comparatively low temperature (150-170 deg C) it loses three-quarters of the combined water and forms a material approximating in composition to the formula $CaSO_4\ ^1/_2H_2O$ and known as plaster of paris or calcium sulphate hemihydrate.

As the product is sometimes used as a gauging material for lime plaster it is often termed gauging plaster. A characteristic of the product is that when it is mixed with water it sets quickly - within a matter of minutes - to a hard mass. This wetting period is too short to permit gypsum to be used satisfactorily as a plastering material by means of conventional trowelling methods of application.

The setting time, as the interval between mixing with water and hardening is called, may be lengthened by the addition of quite small quantities of glue-like materials and the material then becomes a satisfactory plastering material. Such additions are known as retarders and the amount used is adjusted to suit the purpose of the plaster.

The manufacturer's recommendations for the use of any particular brand of plaster should be followed. Simple precautions to be taken include storing in a dry place before use to avoid the possibility of staleness and perhaps flash setting.

Calcium sulphate plasters in general are not suitable for use in damp conditions because they will slowly dissolve in water.

Lime Gypsum Plasters

Mixes based on lime but incorporating a gypsum plaster have good working qualities and reasonable early strength proportionate to the amount of gypsum plaster added.

Gypsum plasters expand on setting and so tend to restrain the drying shrinkage of the lime. For this reason the intervals between the application of successive coats may be shortened.

Fibrous Plaster

Fibrous plaster is a factory product approximately 10 mm thick. It incorporates sisal fibres and gypsum plaster in sheets and moulded forms which can be transported to a building site and fixed to timber or other framing. joints are 'set' with setting plaster using the traditional trowel techniques.

Plaster Boards

Gypsum plaster is now widely used as the core of sheets that are heavily paper covered on both faces and have a very smooth surface. These sheets can be glued or nail fixed to timber or metal framing and can be used to build up a fire resistance rating in partitions and walls. Joints are easily 'set' as above, thus eliminating the 'wet' characteristics of traditional plastering.

Plaster boards have now been developed with water resistant coatings which enable them to be used in bathrooms etc. Wall tiling can be fixed to them using adhesives developed for this situation. These boards usually have a coloured surface to distinguish them from the normal plasterboards. Suppliers should have full technical information.

Rough side of fibrous plaster sheeting showing the embedded sisal fibres which give the sheet its strength.

Casting benches for fibrous plaster cornices.

Cement and fine aggregate mixes are frequently used as a surface finish on paved areas where the base may consist of concrete, brickwork or similar. Such finishes are referred to as screeds or toppings and vary from 10 mm to say 50 mm and are placed in a plastic state. Such toppings can provide necessary falls for drainage on roofs, smooth, slip-resistant, or hard-wearing compounds.

The fine aggregates frequently incorporate sand and a proportion of blue metal dust and screenings which give hard-wearing properties.

Because of the shrinkage of such mixes on drying, they have to be laid in limited areas at one time and joints allowed for.

Such topped pavements depend on the cementitious bond with the base to prevent them becoming 'drummy', so that a roughish base surface is generally desirable. Where heavy traffic is to be encountered, it is usually best to avoid cement-topping screeds, but in some industrial situations they have been found the most practicable.

Self levelling floor toppings have been recently developed which can be poured and swept into place in a semi liquid state, to give a very good level surface when dry.

SUMMARY OF MORTARS, PLASTERS AND RENDERS

Structural

Mortars have compressive strengths which vary widely with the components used.

Cement mortars have higher compressive and adhesive strength than lime mortars. Cement content should be kept to a minimum consistent with intended location, and compressive stresses anticipated.

Plasters and renders are not included in consideration of structural strength of components to which they are applied.

Manufacture

Lime and cement are now produced in factories and the product transported to the building site, usually in bagged form.

Sands are excavated, washed and graded as a mining operation to supply the needs of cities, but clean local sands from streams can be used when readily available.

Lime mortars are usually mixed at a factory and truck delivered, but may still be site mixed.

Cement is measured and added to mortars and renders at the building site shortly before placement. Mixing is sometimes done using a hoe-like tool called a '**larry**' or in a rotating batch mixer.

Water

Mortars, plasters and renders involve 'wet' trades which use water to give plasticity to the materials and generate chemical and physical reactions. Absorption of water from the plastic mix into the wall is an important function in generating adhesion between the two materials.

Freezing conditions before drying is complete can seriously affect the quality of finished plastering.

Rapid drying, due to heat and winds, can cause cracking and loss of adhesion of applied plastering.

Cement content causes shrinkage with drying and fine surface crazing often becomes apparent without indicating defective work.

Salt solutions can cause deterioration to mortars and plasters and affect adhesion. These conditions are often associated with rising damp.

Gypsum plasters are not water resistant and must only be used internally and in dry areas.

Heat

Atmospheric temperature variations do not seriously affect mortars, plasters and renders.

Fire can cause lime plasters to lose adhesion to their base wall. Cement renders are considered by some authorities to increase the fire-resistance rating of brick and concrete walls.

Gypsum plasters, sheets and boards are fire resistant. When applied in accordance with local codes, they can be used to achieve fire-resistance rating.

Ultra-Violet Radiation

Lime and cement products are not affected by sunlight exposure.

CEMENT SHEETS

Electrolytic and other Characteristics

These products are non-conductive of electricity.

Some slow reactions occur between lime and lead over long periods of time.

Vibration during setting is damaging to lime plasters.

Lime plaster surfaces are easily scratched and not as hard wearing as cement plasters and renders.

Acoustic Properties

For effective soundproofing of masonry walls, the joints need to be fully grouted up. Plasters and renders can add to sound insulation.

The shapes and surface finishes of plaster work can be an important consideration on internal acoustics of spaces by influencing the reflectance, absorption, and consequently reverberation times.

Many special acoustic ceilings and wall elements incorporate perforated cast plaster products.

CEMENT SHEET PRODUCTS Fibre-Reinforced Cement

The practice of using hair and other fibrous materials to reinforce plaster of paris and clay products is quite an old one in the building industry. It has been observed in some very ancient sundried bricks and similar materials in early civilisations and succeeding generations ever since.

In the 20th century sisal fibres have been used to give tensile strength to plaster sheets known as fibrous plaster and to strengthen building papers. Asbestos fibres have also been used widely with cement and fine sands to produce asbestos cement in a great variety of forms, but asbestos fibres are now banned in Australian building products because of their danger to human health. If asbestos cement products are found on a job, do not attempt to cut with power tools, such as angle grinders. Special precautions are necessary. Many other types of strong fibres are now substituted, and the range of products has expanded. See James Hardie website: www.jameshardie.com.au.

The basic strength of fibre-reinforced cement products is derived from the large number of overlapping fibres incorporated into the products (so that some tensile strength is developed in all directions) combined with the hardness of the plaster of paris or cement base able to resist compression. Quite thin sheets (e.g. 3 mm) can be produced with a high degree of stability in this way.

A cement rendered wall screens the entrance to a school building which uses fibre-cement flat and moulded sheets of 12 mm thickness for external cladding.

Asbestos Cement (Fibro)

Asbestos fibres have been used since early the 20th century as reinforcement for thin sheeting materials both flat or formed into a variety of shapes. The resultant sheeting was hard, inert and weatherproof and suited to wall and roof cladding. In a corrugated form it was used extensively for roofing many buildings, especially factories. Its major defects were the brittleness which developed with age and low insulation values for both heat and acoustics. It also tended to explode in fires.

Many components such as gutters, pipes, rainwater heads, etc. were made from this material.

During the 1960s serious concern developed regarding health aspects of the mining and manufacture of asbestos products which led to a ban on their use in many materials and particularly where applied as a sprayed-on form for fire protection in buildings.

In Australia the manufacture of asbestos cement is now phased out, but some countries are still mining asbestos and making these products.

Other fibres have now been developed to replace asbestos. While the general whitish-grey colour and appearance of the products have not changed greatly, the qualities of inflexibility and brittleness have been markedly altered and a greater range of shapes and surface patterns has improved market acceptance.

Fibre-Cement

Modern fibre-cement products are the outcome of the changeover from asbestos to other fibres as cement reinforcements. Earlier the term fibro-cement or 'fibro' was commonly used in Australia for asbestos-cement, but the modern product is far superior in flexibility, quality and performance to the old 'fibro', but almost indistinguishable to casual observation. One of the main fibres used is cellulose.

A wide range of building components is now produced as standard items. Many buildings have had wall and window components specially formed as one unit. Techniques of fixing and jointing have been greatly improved so that some fibre-cement clad walls of buildings now reflect high technology concepts and performance. On the other hand, planking to simulate traditional weatherboards has proved very popular and has the benefits of easy fixing, fire resistance and minimal maintenance. These planks are available plain or suitably patterned.

Glass-Reinforced Cement

Research and sponsorship by the Pilkington Glass Co. in Britain has led to the recent development of glass fibres able to resist alkali attack and resultant glass-reinforced cement products. Care has been exercised through a strict licensing system to develop a high-quality product. So far components have not been of a major structural nature, but results suggest that there is a very wide range of application for G.R.C. in buildings.

The tensile strength developed by glass fibres is extremely high, far in excess of comparable steel fibres. G.R.C. fabrication has been well established in England and Europe. The first significant use on a public building in Australia was for Telecom in Brisbane, where its potential with moulded forms was well displayed.

By use of this material, much thinner sections are possible than with reinforced concrete or even ferro cemento, and hard, sharp, durable surface finishes are available. As no steel is used, durability of the product is not threatened by corrosion of steel reinforcement. G.R.C. can be produced using short fibres in random form impregnated with the plastic cement. Gun spraying techniques are used to mix and apply the material onto moulds. See GLASS for illustration on page 84.

Casings for sandwich-type panels, which incorporate other lighter insulating materials between outer layers, have already proved successful. Street furniture and decorative moulded panels are other obvious adaptations.

G.R.C. plaster on concrete block walls, laid up dry without mortar, can provide waterproofing and structural properties similar to those of a conventionally laid wall. This is an area in which much development work of new techniques is under way and G.R.C. promises to be one of the materials of the future.

PLASTERS AND RENDERS

Standards

AS 1639-1990	*Fibre reinforced cement roofing & cladding*
AS 1672-1974	*Building limes*
AS 2185-1978	*Fibrous plaster products*
AS 2186-1978	*Code for erection & fixing FP products*
AS 2588-1983	*Gypsum plasterboard*
AS/NZS 2589.1:1987	*Gypsum Linings*
AS 2590-1983	*Glass fibre gypsum plaster*
AS 2591-1983	*Fixing glass fibre gypsum products*
AS 2592-1983	*Gypsum plaster for building*
AS.CA27-1959	*Code for internal plastering on solid backgrounds*
AS 2908-1992	*Cellulose cement products*
AS 3582-1991	*Supplementary materials for use with Portland cement*
AS 3792-1991	*Portland & blended cements*

CSIRO Publications

NSB 66	*Wallboards in fire ,*
NSB 70	*Mortars for masonry*
NSB 71	*Plaster mixes*
NSB 87	*Fire resisting materials*
NSB 98	*Light-weight construction*
NSB 165	*Wall cladding*
NSB 175	*Toppings for concrete*

A light cement render coat such as this is not thick enough for a smooth finish but provides a pleasing texture or base for further finishes.

Plain and decorated perforated plaster ventilators for internal use.

Perforated plaster wall used for acoustical control in courtroom. Note also use of decorative plys and laminated timbers.

STONES
Rock & stone; quarrying & shaping; stone for concrete; commonly used
building stones, decay; references; characteristics summary; stone products.

INTRODUCTION
In most areas of civilisation where stone is available, it has been a favoured material for buildings of importance and permanence for up to 5000 years. The qualities of stones vary, depending on their geological origins, but most used for building purposes are basically fire and weather resistant and not affected by insect or organic attack; consequently many of the great historic monuments of the past are constructed largely of stone.

Stone's good compressive strength allows it to be used for walls and arches, but most stones lack sufficient tensile strength to be used for long-span beams.

Being naturally produced materials, stones cannot be guaranteed to meet the same established performance standards as factory-made products, so sampling, testing, observations and traditions have to be taken into account to select and use stone wisely. This fact is becoming increasingly important now these materials are used as veneers to achieve decorative finishes rather than as solid structural members.

In these thin-member situations differential movements, thermal effects, weathering, etc. become much more critical than in solid masonry construction.

Stones can provide beautiful colours, textures and effects in buildings, and weathering qualities are far better for many stones than for modern precast materials. It is unlikely therefore that natural stones will be superseded completely as surface finishes.

The crushing strength of a stone is not a criterion of its quality or durability. Crushing strength is affected largely by moisture content and whether pressure is applied parallel or perpendicular to the bedding planes.

Rock and stone relationships
The three major categories of rock from a building point of view are igneous, sedimentary and metamorphic. Each of these has a basically different geological history which affects its appearance and performance in building.

The igneous rocks are primary rocks and come from the molten magma of the earth's core which solidified when it approached or reached the crust of the earth. These rocks solidify in a crystalline form which is uniformly hard and usually without defined layering or veins. The granites, basalts and trachytes belong to this group.

The sedimentary rocks are secondary rock formations created by sedimentation of geological and/or organic materials usually deposited in water in successive layers and gradually compressed into solid rock formations.

The major sedimentary material in the rock influences its name but is not always a true guide - e.g. sandstone is composed largely of quartz grains, limestone usually includes marine life deposits such as coral or shellfish. In many of these stones the layering effect is clearly marked in the stone when it is cut. Fossils are frequently found in limestone deposits.

The metamorphic rocks might be termed tertiary rocks because they are often the product of some erosion or sedimentary process plus the influence of heat, extra pressure, etc. Many marbles are metamorphosed limestone, but some stones are called marbles even though they are really limestone with a strong grain or pattern.

The extra heat or pressure usually gives the stone additional strength, hardness or colour.

Quarrying and shaping
All stone is initially won from the earth's crust by some form of quarrying. Methods of quarrying vary accordingly to the type and hardness of the stone, its natural bedding and planes of weakness. Most rocks are compressed, shrunk and stretched in their geological formation, causing real or incipient parting planes.

For a quarry to be viable commercially the stone deposits need to be extensive, usually deep and of consistent quality.

Although stone underlies most of the earth's crust much of it is too irregular or of too doubtful a quality to be used as building stone. Sydney, for instance, is located on a huge sandstone formation, yet great difficulty is being experienced locating a suitable new quarry site needed to maintain and restore the many old sandstone structures. The quarries previously used have now gone out of production and their sites converted to other uses.

In some parts of the world such as the Carrara marble quarries near Pisa in northern Italy, the huge rock walls have been quarried for many centuries.

To remove the stone for building blocks requires slow and careful drilling of holes and insertion of wedges to split the rock into large blocks in the required places. Some modern developments include use of gas and water under pressure, to force the rocks apart. Previously hand-hammered metal wedges or timber wedges, swollen by watering, often provided the pressures needed.

The initial large blocks are moved to cutting sheds where mechanical saws are employed to break the stone down to usable sizes. Sand and water, carborundum grit or wheels of hard alloys and industrial diamond saws are all used in the shaping processes, dependent on the type of stone and equipment preferred or available.

Finishing and polishing require another set of machines and abrasives to produce the surfaces specified.

Hand chiselling is still used for some fine detail work, especially where restoration or renovations are involved.

These hand methods have changed very little over centuries of use. With them a great variety of surface textures, patterns and ornament are possible.

Stone turning is also used to produce rounded balusters, etc. as seen in some illustrations.

Stone for concrete
In this century the quantities of stone quarried for concrete manufacture and other industrial uses far exceed that for other building purposes. These usually need the material in easily handled bulk form such as small pieces of stone. River pebbles are one source of supply for this market, but the major needs are met by use of crushed stone.

As the size of stones can be readily sieved, explosives or rippers can be used in quarrying, as the shattering effect which would harm dimensioned stone is not important.

Concrete aggregates are best made from granitic or basaltic stones, which have good compressive strength, are not readily affected by water and crush to a hard, sharp-edged form which bonds well with cement and sand. Concrete mixing plants usually secure well-tested quality material for their purposes to enable their specified mixes to be maintained under field test conditions on buildings.

For other purposes such as road building, some different aggregates may be required because of specific qualities needed to withstand modern traffic conditions.

COMMONLY USED BUILDING STONES
Granites are coarsely crystalline stones which range in colour from almost black through greys, browns, reds to pinks and yellows. The reddish-brown colours are usually due to iron stainers or feldspars. There are often two distinct coloured crystals in granite specimens, one being white quartz. The proportion of white to coloured determines the apparent colour of the specimen.

Granites are hard wearing and can be polished to reveal the full colour and texture of the stone. They have a relatively high tensile strength for stones and are among the heaviest of these natural materials, with high compressive strength; and they are non-porous.

A monumental building using base of solid igneous stonework, marble column claddings and travertine wall facings.

Basalt, often called bluestone, is generally a finely grained crystalline stone of dark grey to blue-grey, sometimes with white quartz seams in it. It is the stone generally preferred for crushed aggregate for concrete when available. It comes from volcanic origins and is closely related to other volcanic rocks called diorite, gabbro, lavas and volcanic tuffs, which, however, may be comparatively less dense.

Basalt is not usually polished, but it can be. It is very hard wearing and for construction is similar to granite in most of its qualities.

Basaltic fieldstones used in walling.

Trachyte is very similar to basalt in texture but of a khakibrown colour. It is hard wearing and is used widely for central Sydney kerbstones where heavy pedestrian traffic is experienced. Occasional surface chipping keeps the kerb from becoming glass smooth from wear.

The stone was also popular for masonry work in Sydney buildings prior to 1920, and some examples show it in several different surface finishes ranging from rock faced to polished.

The source of most trachytes used in Sydney was the Bathurst area, over a hundred miles inland.

A working rock face in a hard stone quarry. Note the varying colours of rock available from this one location.

Polished red and smooth faced grey granites after 50 years exposure.

General view of crushing plants and stockpiles of graded aggregates in a large hard quarry operation.

Solid trachyle in rock faced, smooth and carved finishes.

Sandstones consist of sand grains cemented with silica, calcium carbonate or clay. The cementing material greatly affects the stones' durability when exposed to weathering. Colours range from white through creams and pale browns to purples and orange. Sometimes strong bands of colour occur within lighter stones due to the presence of iron solutions.

Hardness and granule size can vary greatly. Most sandstones are porous and therefore affected by moisture, pollution or foundation dampness.

Many sandstones show clear bedding planes in the quarry. These stones can often be split easily for use as paving stones or coursed rubble walls.

Freestone is a thick-bedded sandstone free of obvious cleavage planes, so it does not tend to split in any direction. It can be cut into blocks of any size and form. Most Sydney sandstone buildings used local freestone.

Foot traffic and running water will erode sandstone, but many kerbstones have survived suburban conditions for 70 years or more.

Minimum dimensions for veneers should not be less than 50 mm thickness.

Fine surface finishes are not currently advisable in sandstone, although some carved work has survived for more than 100 years. Smooth surfaces are commonly used, but no polishing is possible.

Sydney sandstone originally came from quarries close to the city centre (Pyrmont especially), but Hawkesbury sandstone from the Gosford area north of the city became the usual source by the 1930s.

Sandstones are universally used and the variations in texture, colour and durability change from quarry to quarry. For this reason it is often very difficult to obtain matching stone to repair old buildings if the original quarry has ceased production. Reworking re-used stones is not always effective.

Traditional handworking of sandstone for restoration work.

These decorative sandstone details are now over 100 years old and in good condition.

Limestones are the most widely used of all the building stones throughout the world. They can vary greatly in hardness, quality and durability but are usually of a white-cream or grey colour. They are slowly soluble in water and very reactive to most acids because a major component is calcium carbonate. Limestone can readily be finished to a soft, lustrous and smooth surface, but it will not take a high polish.

Limestones frequently have local names such as alabaster, etc. which should be used to distinguish a particular choice if needed. Compressive strength is usually adequate for building stresses in masonry walls where stones of 200 mm plus thicknesses are used.

As foot traffic or running water will erode this stone surface its use should be confined to walling situations.

Some limestones can be very easily worked and carved so that highly decorated details are possible.

Some very soft and easily worked forms of limestones harden on exposure. Extensive use of such stone was made in the construction of Adelaide, much of which came from the Mt. Lofty ranges.

Travertine, although containing many voids of quite large dimensions, is now exported and used in many countries of the world. The voids are the result of the natural formation of the material in a hot spring area of Italy, and holes often measure 50 mm x 15 mm on the surface.

Travertine wall showing the open holes common to this stone.

STONES

47

Portland stone which features in many London buildings is also a limestone, of pale grey colour and fine texture. It is the similarity to this stone that generated the name 'Portland cement'.

Marbles are metamorphic stones usually more colourful and noticeably patterned than the sedimentary group and are quarried in many locations throughout the world. Because of the massive deposits of rock and their highly regarded qualities, Italian marbles are exported to many countries and have dominated the Australian market for many years.

There are, however, some excellent Australian marbles and the recent extensive use of Wombeyan marble in the new CBC Bank in Sydney was probably the first such major installation for more than 40 years. This softly patterned marble has a beautiful golden cream colour which enables it to be used in large areas without being too dominating.

The colour range of marble is almost limitless, but stocks maintained have to be restricted so that advance ordering is necessary for large projects, to allow time for quarrying, cutting, polishing, etc.

When used for veneer, marble needs to be approximately 50 mm thick externally in order to retain stability. The jointing and fixing must be carefully considered in relation to the likely expansion, deflections and creep in the structure. Water entry behind veneers can be disastrous if freezing occurs.

Marble can be carved or turned and brought to a high polish. It can be used as paving slabs, stair treads, balustrades, wall panels, column encasings, etc. and many examples of its use are seen in most cities.

In selecting marble the recommendations of the suppliers should be sought regarding the suitability of a particular stone for the purpose intended.

Marble floor and casings to columns with polished granite cantilevered seat in foreground

A recent interior use of Australian Wombeyan marble facings of golden tone.

Slate is usually dark grey with bluish to purple tones. It can be split into very thin sheets - as for roofing slates.

Slate is also wear resistant as step treads and paving.

Most slates are imported from Wales or India, but local slates are available in South Australia, and in Adelaide some very large slabs are used as city pavements.

Before metal damp courses became readily available slate was often used in two or three layers to form the damp-proof course in solid masonry walls.

Slate door thresholds and step treads were popular in buildings up to 50 years ago, usually being approximately 30 mm thick, with a rounded front edge.

Slate roofing which was popular in 19th and early 20th century is still available, but heavy and usually imported. Slate roofs lost popularity from 1915 when Wunderlich commenced local manufacture of Marseilles pattern terracotta tiles.

STONES

DECAY OF STONEWORK

Despite the hardness and strength of many building stones, all are subject to decay, some more rapidly than others.

High levels of air pollution in modern cities contribute greatly to these problems, as many damaging acids and compounds are deposited on buildings and attack the stones or their bedding material. Sulphuric acid, carbon dioxide, chlorine, ammonia compounds, sea salts, etc. are all dangerous and are found in many city atmospheres.

Water, even if pure, can be damaging, especially if in any concentrated form. High-pressure water or steam should generally be avoided for cleaning purposes because of the potential danger to stone facings.

Wind can seriously damage soft stonework, especially where it carries grit, dust or decomposing vegetable matter.

All stonework should be subjected to regular cleaning by water soaking and gentle scrubbing to remove collections of deleterious matter. Use of chemically based preparations for cleaning need great care and expert application, and should generally be avoided. Maintenance and repair of old stone buildings has become a major activity for many organisations and a revival of interest and activity in stonework is apparent in many countries.

Old graveyards give excellent comparisons of relative weathering properties of different stones.

Fine jointed ashlar masonry showing differing deterioration of sandstone in adjacent blocks.

SUMMARY OF BUILDING STONES

1. Structural strength

Most stones are capable of resisting compressive stresses encountered in normal walling situations, but stone for columns and other components of concentrated load should be carefully selected.

The crushing strength of stone is not an indication of its quality or durability and can be considerably reduced if the stone is saturated.

Tensile stress resistance of stones is generally considered to be minimal and not used in stress calculations; but some igneous rocks do have reasonable tensile strength and consequently shear resistance.

2. Manufacture

Quarrying of dimensioned stone is now restricted to very few locations and long haulage may be necessary for some types. Initial cutting and shaping is often done at the quarry with further polishing and finishing at the supplier's yard.

Work on the building site should be kept to a minimum and any finely finished surfaces, arises, etc. need to be carefully protected from damage in transit and on site.

3. Water and freeze/thaw effects

Porosity of stones varies greatly with the igneous and metamorphic stones usually being less porous than the sedimentary group. Limestone is slightly soluble in water.

Where water can enter stonework either at the joints or via porous surfaces, it is likely to cause damage and disfigurement. If the area is subject to freezing, cracking and spalling on the surface can become serious.

The well-respected building stones are usually not seriously porous and can make waterproof solid walls.

Water is the least damaging aid to clean stonework of pollution and stains.

4. Temperature effects

Normal seasonal variations of temperature seem to have little effect on solid stone walls, and stone masonry is fire resistant, but differential movements between stone veneers and backing materials can be dangerous.

Fire can be damaging, especially to limestones. Stone veneers will usually crack and possibly become detached from their supports in a fire, especially if hosed while hot.

Some stones can explode in fire and care must be taken in selecting stones for fireplaces in order to avoid such problems. The heat absorption and storage capacity of stone makes its use desirable where dramatic climatic thermal effects need to be reduced for interior comfort conditions. Thick stone walls can absorb a great quantity of heat, especially from direct sunlight, then re-radiate the heat when temperatures drop. Conversely, stone buildings usually remain cool while external temperatures climb on a hot day, but may still be warm into the night.

The use of stone as a heat sink in buildings designed to maximise solar benefits is becoming popular.

5. Ultra-violet radiation

Most natural stones are unaffected visually by the impact of ultra-violet light. The colours of natural stones are remarkably permanent, especially if kept clean.

6. Electrolytic or other special effects

Stones are non-conductors of electricity and generally are nonreactive with other materials.

Some combinations of stones - e.g. limestone above sandstone where exposed to weather - can lead to staining and accelerated deterioration of the lower stonework.

Very soft stones can be eroded by continuous strong winds.

7. Acoustic qualities

Being of solid, heavy material, stone walls usually contribute to highly reverberant internal conditions. They are also good sound insulators as partitions between spaces.

The surface finishes used will affect acoustic qualities, polished surfaces being more reflective of sound as well as of light.

References

Because of the variable nature of stones, local references are best consulted for local materials, but the importation of stones has rendered some information of value internationally. `A Technical Guide to the Rational Use of Marble' published by the Italian Marble Industrie 1972 is probably the best book on modern practice.

Most other books on stones are products of the age of solid stone construction prior to 1920 and may be found only in major libraries.

The Australia/New Zealand Building Construction Materials and Equipment Number 151, Oct./Nov. 1983 contains a very useful article on various stones by George Mills, which includes information on surface finishes.

Papers produced by Dr Ian Wallace, Principal Geologist, NSW Department of Mineral Resources and Development, are useful references, especially regarding modern quarrying and stoneworking methods. His paper, Stone, published as part of the 1979 Traditional Building Technology Seminar at the University of New South Wales Graduate School of the Built Environment, lists local quarry locations and illustrates techniques now used.

The National Trust (NSW) publication, Maintaining & Restoring Masonry Walls, Gibbons (1978) records seminar information of 20th century methods in use in Australia.

AS 2758   Aggregates & rock for engineering
NSB 57   joints in masonry
NSB 62   Masonry veneer construction
NSB 65   Structural masonry applications
NSB 76   Silicone water repellants
NSB 111   Stone veneer fixing
NSB 141   Bond strength in masonry
Local stone suppliers should be consulted for advice.

A terrazzo tile pavement showing pattern of contrasting colours.

Pebble finish as applied to many surfaces.

STONE PRODUCTS

Generally

Stone is now used in numerous products incorporated in building work, especially for decorative purposes in floors, exterior walls, steps, paving, etc. The uses of concrete, which is often described as artificial stone, are dealt with separately.

Many of the stone products are closely related to concrete in methods of manufacture but are specially treated to display the natural qualities of stones used as aggregates.

These products can generally be produced and applied at lower costs than can traditional stone finishes. Their permanence is not proven as well as natural stone's. Care must be taken in making choices, especially in exposed and important locations.

Terrazzo

Terrazzo has been used since Roman times. Essentially it consists of marble chips placed in a cement or a non-cement matrix which is then ground down to a smooth, polished surface, or is given some other finish after curing. It can be placed in situ or in precast form.

Terrazzo offers a very wide choice of materials, and therefore of effects. Marble chips can be selected from a considerable range of colours, textures and types, whilst the chip sizes themselves can range from 2 mm to 150 mm in diameter. Small chips of 3-10 mm are commonly used but larger pieces can achieve special effects. Precast terrazzo tiles are widely used for floor surfacing.

Special additives, such as carbon black, can give conductive qualities necessary for such locations as hospital operating theatres.

Epoxy terrazzo involves the use of epoxy resin instead of the cement and can be laid very thin and provides protection against spillage of fats and mild acids. In addition it provides better soundproofing than normal terrazzo.

Methods of laying terrazzo need to be carefully considered in consultation with the manufacturers to avoid cracking or differential movement with the structural floor. Precast slabs are also used, often reinforced and in 37 mm thickness, as WC partitions or as facings.

Pebble and crushed aggregate facings

These facings are now widely used, especially in association with precast cladding slabs or reinforced concrete.

The facing stones are selected for colour and texture. Sometimes they are placed in the formwork before the concrete is poured, or, in other methods, applied soon afterwards while the concrete is still fresh, producing a veneer integral with the main slab.

After curing, the surface materials are exposed by brushing away some of the cement and sand between the stones. This process is usually assisted by the use of setting retarders or inhibitors on the finished surface; but care has to be taken to ensure adequate adhesion to the structural member.

Veneer-type panels on some prominent buildings have shown weaknesses and separation from the parent body after a few years and therefore appear to be suspect regarding their permanence.

Coarse and fine crushed aggregate surfaces on precast concrete panels.

Reconstructed stone

This method employs normal concrete techniques but includes only aggregates selected for their size and colour. Before completion of the work the surface colour is developed by cleaning back with grinding, sandblasting or acid etching.

With this technique a variety of shapes can be cast incorporating the structural benefits of reinforced concrete work and the colours of selected aggregates.

Sydney Opera House displays great quantities of reconstructed pink granite in the podium area, and it appears to be weathering well.

A less successful earlier use was on a large fountain where the water very quickly caused erosion and blockage of the pump mechanism. In this case solid granite was effective as the replacement.

For many situations, however, reconstructed stone has been used effectively in conjunction with natural stone, especially where repetitive carved detail work would be very expensive.

Modern machining equipment and methods have enabled a variety of polished, sandblasted, or other finishes to be offered at competitive rates.

Reconstructed pink granite steps at Sydney Opera House - when over ten years old.

A 19th-century building showing the compatibility of brick, stone and ceramics as masonry materials.

References

NSB 62   *Masonry veneer*
NSB 109   *Terrazzo*
BTF 20   *Specification of Dimensioned Stone*

Building Construction Materials & Equipment magazine Feb-Mar `83 should be consulted for Terrazzo; its installation & maintenance.

CERAMICS

History; manufacture; glazes; terms. Building products; terra cotta tiles; blocks, pipes, sanitaryware; floor and wall tiles, references; summary.

INTRODUCTION

By 'ceramics' is meant all production that results in baked clay of different grades of hardness and purity.

The term 'pottery' is also used widely to categorise these products. Factories and kilns devoted to this type of production are often called 'potteries'.

The long history and usefulness of ceramics derive from:

(a)   the widespread availability of clays as raw materials;

(b)   the plasticity of clay when wet;

(c)   the conversion of clay into an almost indestructible material by firing at high temperatures, and

(d)   the ability to apply permanent colour and pattern to the products by glazing.

History

Ceramics, like bricks, are among the oldest of man's manufactured artefacts. They are made from clays, generally of finer texture and lighter colour than are used for bricks; and over the centuries the range and size of articles has become very extensive. Now it is possible to buy fine drinking cups or heavy sanitary pans; sewerage pipes or beautiful vases; fine mosaics or heavy roofing tiles - all products of the worldwide ceramics industries.

In building, the history of ceramics is closely related to the use of bricks and tiles for roofs and the decoration of important features.

The early Mesopotamian builders used bricks with glazed faces of various colours. The early Mediterranean and Chinese civilisations all used terra cotta tiled roofs. Hollow pottery components in walls or dome-type elements have been used to reduce dead weight for structural reasons. Earthenware pipes for drainage have been found in archaeological excavations. Decorative glazed tiles and porcelain have come to us from techniques and developments in China and the Islamic civilisations and were first manufactured in Europe about the 15th century.

In Australia, Fowler's Potteries commenced producing earthenware pots and pipes at Camperdown (Sydney) in 1837, but other small kilns may have existed before that time.

In 1915, World War I caused restrictions on imports of the popular French (Marseilles) pattern tiles, so the importer, Wunderlich, then opened a factory for roofing tiles at Rosehill, west of Sydney. Most of the finer quality floor and wall tiles were still imported from England until large-scale ceramic production commenced at various local factories in the 1950s and 1960s.

Australia now has available a very wide range of wall and floor tiles that come from many industrialised countries, as well as from local firms, including modern wall tiles for building interiors. Where colour is not a critical factor, many ceramics are made of coarse-grained and coloured clays that occur widely throughout the world.

MANUFACTURING PROCESSES

The unglazed products are usually the result of one hightemperature firing, whilst most of the glazed products require an initial firing to produce the 'biscuit' to which the glaze is then applied and the product re-fired. Some of the multi-coloured ceramic products, particularly the porcelains, are fired three times before the finished result is achieved.

Traditionally, the range of colours that could withstand the heat reliably was restricted, but modern chemical engineering and quality controls have produced a very extensive variety of colours and finishes. Gas, oil and electrically fired kilns now create the accurate temperatures needed for consistent results.

GLAZES

Materials used for glazes are many and varied, but the most important kinds of glazes are as follows:

(a) **Alkaline (soda-lime and other materials)** - the most ancient and uncertain, used by ancient Egyptians, Syrians, Persians.

(b) **Lead** - the most widespread in use and the best for all ordinary purposes.

(c) **Feldspathic** - the glazes of hard-fired porcelain, generally unsuited to other materials.

(d) **Salt** - produced by vapours of common salt; the special glaze of stoneware.

Some glazes are placed on articles by dipping, spraying or brushing, whilst others are produced by vapours created within the furnace. The variations and complexities are far too numerous for study in a course such as this on building materials.

CLARIFICATION OF TERMS

Some common terms associated with ceramics need clarification, because they are often very loosely used, e.g. -

Terra Cotta - A fairly coarse porous type of product, dull ochre to red in colour, which can be glazed or left unglazed. Has been used in buildings for many centuries before Christ.

Earthenware - All clay articles which, when sufficiently fired for practical use still remain porous and opaque, are classified as earthenware. These may be glazed or unglazed.

Stoneware - Basically of the same constituents as terra cotta but impermeable, having developed more hardness from firing. May be glazed or unglazed.

China - The fine white glazed household ware which originally derived from the porcelains and white glazes of Chinese civilisations via the Islamic copyists. They were not produced in Europe until the 15th century.

Porcelain - The fine translucent white glazed ware originated in China, which uses feldspathic material for glazing.

CERAMICS IN BUILDINGS

Only a small proportion of total ceramic production is devoted to the building industry, but some factories do specialise in building products. The major types produced are described as follows:

Terra Cotta Products

The variety of terra cotta products used in building over the centuries has been tremendous. Local examples often can be seen of elaborate ornamental detail incorporated with brickwork. More utilitarian uses are for air bricks (ventilators), hollow brickwork in walls, and/or reinforced concrete-ribbed slabs, to minimise dead loads.

By far the commonest use, however, has been for roofing tiles in many parts of the world, especially in the Mediterranean countries, Australia, Asia and parts of Latin America.

A small selection of some ceramic products stacked in a supplier's yard.

Terra Cotta Roof Tiles

These roof tiles, although brittle, are very permanent in resisting most temperate to hot weather conditions and are generally of the typical reddish tone of terra cotta. By the addition of glazes to the red biscuit, colour variations are possible, and the tendency to water absorption is minimised.

The range of patterns varies with local practice and factories, but historically the basic patterns are S-shaped pan tiles and the U-shaped Roman tiles, in both of which there is a clear ridge and gutter pattern for channelling water run-off. This moulded form helps to give the tiles structural strength and allows for some irregularity in the overlaps required to adjust to varying building conditions.

The popular French or Marseilles pattern tiles were developed in the 19th century and first imported to Australia about 1890, where they proved more suited to the environment of the populated coastal regions than did galvanised iron.

Variations of the basic pattern are numerous, but the general size of Modern roofing tiles in Australia is fairly uniform at approximately 428 x 254 mm. A range of ridge tiles and accessories is available, although now not as varied or decorative as when these products were imported.

Other smaller-sized tiles or 'shingles' are sometimes used but are more expensive and not always readily available.

Most of the locally-produced tiles are now glazed. The only serious disadvantages with these tiles are reduced life in positions exposed to salt-laden atmospheres, and their brittleness if subjected to hard knocks and impact from heavy hailstones.

A terra cotta tiled roof using a dark manganese glaze.

Glazed Marseilles pattern terra cotta roof tiles.

Finely detailed faience work executed about .1930. Now in a highly polluted atmosphere and showing deterioration.

Glazed Terra Cotta Blocks (Faience)

In the latter part of the 19th century a revival of interest in glazed terra cotta as an external facing material for buildings developed, largely to combat the effect of pollution on natural stonework which was becoming serious at that time in England and in other industrial centres.

Two facings developed:
(a)   the glazed hollow terra cotta block
(b)   the glazed solid terra cotta slab

Each of these materials is designed to be a facing only to masonry walling, to which it is applied.

The blocks are designed to avoid sharp corners where chipping of the glaze is likely to occur. They are usually located away from doorways and likely damage points. Studies undertaken in Britain in 1928 showed the blocks in use to be highly resistant to atmospheric effects in 50 years' exposure.

The hollow blocks are usually made to be laid as brickwork or stonework on a 115 or 230 mm bed. Sizes and detailed shapes on facework varied to suit manufacturers' and designers' requirements, with a maximum face size of approximately 600 x 300 mm. As considerable shrinkage occurs in firing, the hollows assist in controlling the evenness and final shape.

Many important public buildings and offices built in the period between World War I and World War II incorporated this type of facing material. It was made locally by Wunderlich's factories at Sydney and Melbourne and is now out of production in most parts of the world.

Some buildings have had considerable difficulty with such facings, but this appears to be due to faulty installation rather than defects in the product itself.

Renovation of such facades may require re-establishment of manufacturing facilities, or removal and replacement, both of which are very expensive; however, the majority of these installations appear to be performing reasonably well.

Glazed Solid Slabs were made as standard items approximately 40 mm thick, with deep dovetail grooves in the back to bed-in with cement mortar back-up. The bedding system often incorporated metal cramps or ties to the masonry backing when the slabs were subjected to traffic vibration. Obviously this slab-type terra cotta creates a much thinner veneer than the hollow block system and is therefore less permanent, but it can more readily be removed and replaced.

For more detailed information on these products and their installation, the old building trade catalogues, handbooks and specifications of the day are very useful references.

Drainage Pipes

Until the 20th Century most underground waste water and sewer pipes from 100 mm diameter to 300 mm diameter were made from saltglazed earthenware pipes of the spigot and socket pattern, and these pipes are still widely used for drainage purposes.

Their major disadvantage was brittleness and inflexibility of the cement jointing system, but use of rubber gasket joints has minimised the latter objection.

Pipes are available in lengths of 600 mm to 1500 mm; and there is a wide range of bends, cleaning eyes, traps, gullies, sinkstones, junctions and other accessories necessary to construct a complete drainage system of ceramics.

Sewerage systems require the best quality stoneware pipes, whilst rainwater drains use the more porous earthenware pipes. Unglazed 'agricultural' pipes, both perforated and plain, are used widely for subsoil drainage in buildings and in agriculture, but these tend to be shorter in length and are not available in such a variety of sizes. They are often straight, without spigots, so that joints are laid 'open' to collect water.

A variety of stoneware pipe junctions, bends, etc.

Sanitaryware

As many items of sanitaryware are ceramics, their production is now carefully controlled, not only by the manufacturer, but also by inspectors representing the local water and sewerage authorities, whose stamp or mark has to be visible on items to be connected to their services.

The glaze on sanitaryware simplifies cleaning and minimises the collection of bacteria, etc.; but it follows that cracked or chipped glazing should not be allowed for such items. Due to slight irregularities of shape sometimes occurring in firing, not all items emerge from the kiln perfect, so that 'seconds' or 'not perfect' products can be obtained which still satisfy all health requirements, but they should not be used where quality is more important than cost.

Some of the commonly used sanitaryware items are handbasins, cleaners' sinks, laboratory sinks, W.C. pans and cisterns, urinals, bidets.

Floor and wall tiles

Floor and wall tiles in the past enjoyed popularity largely due to local artistic skills and technologies, but now, with world-wide transport and trade, the range and availability of these products has increased enormously. They are usually flat and readily packed into easily-handled cartons for site use and larger crates for shipping economically.

Ceramic basins and toilet fittings in a new bathroom which also uses timber, laminate-sheeted chipboard and glass products.

The major advantage of tiles are:
(a)   their permanent colour, pattern, texture;
(b)   the ability of glazed tiles to resist absorption of water, fats, oils and many chemicals;
(c)   their ease of cleaning and durability;
(d)   the way tiles can produce a satisfactory finish when laid on a cement mortar bed over a comparatively rough surface of concrete, brick or stonework.

Some tiles curl or warp during firing, therefore these are usually culled out in the factory and the better ones sold as 'seconds' at reduced prices. For first-quality work, however, the tiles should all be flat and uniform in thickness.

Some special cove tiles for wall to floor junctions are available, made to a wide range of sizes and colours.

Floor tiles of small dimensions, from approximately 10 mm x 3 mm thick, are often assembled into patterns in the factory and mounted on paper or netting in squares of, say, 300 x 300 mm. This assembly assists speedy and accurate laying on the job. For surfaces where floor gradients are variable and much cutting and fitting is necessary (as in bathrooms), these small tiles are often a wise choice.

The range of sizes, shapes, patterns and colours is endless. Some floor tiles are fully glazed and others are matt finished. Traditionally the larger the floor tile the thicker it needed to be to resist cracking. These thick tiles were referred to as quarry tiles. Modern technology can now produce large and thin but extremely strong floor tiles. In selecting floor tiles an important consideration is whether wheeled trolley type traffic will be used. This type of traffic usually requires a thicker tile (possibly concrete) to resist cracking under such loads.

A serious disadvantage with floor tiles is their tendency to be slippery when wet. This has led to the manufacture of special nonslip surface finishes incorporating carborundum granules, or textures, on some tiles. Great care must be taken with the selection of tiles for areas affected by normal rainfall, or water in domestic or industrial situations because of this dangerous characteristic. (Paving bricks are often preferable for external locations because of their rougher texture and less slippery surfaces.)

Floor tiles are usually laid on a thick bed of cement mortar (say, 10 to 40 mm) which is prepared, graded and/or levelled first; then the tiles are set out and patted firmly into the mortar which is laid in an almost dry condition with only sufficient water to create the cement bonding reaction.

Where tiles are to be laid over fresh concrete a special fast curing and flexible cement based mortar bed is needed to adjust to the shrinkage movements which will occur in the floor. A similar bed is desirable on fibre cement flooring panels which have a degree of flexibility sufficient to disturb floor tiles if normal cement & sand bedding is used.

Some tiles have an indentation pattern on the back to assist in producing a secure key between tile and bed. This indentation is usually found on tiles which are very hard and non-porous, because there is no adhesion from suction of cement into the tile.

Wall tiles usually consist of a thin white absorbent biscuit approximately 5 mm thick which is glazed on one surface and slightly textured on the rear.

Glazed tiles, usually square or rectangular, are often used for hygienic purposes on walls, and fine joints only are left between the glazed surfaces to assist with cleaning and maintenance.

CERAMICS

A small display of some floor, nosing and wall tiles.

Where the tiles are placed over a brick or stone wall the bedding and adhesive material is often a cement mortar which assists in producing a smooth, flush surface and good adhesion. Bedding is usually applied to the and the tile is then pressed against the wall. The gaps between tiles are filled later, usually with white cement or a plastic pointing material, to prevent water penetration behind the tiles to the bedding.

Alternately wall tiles are applied to an already cement-rendered wall face, using a modern flexible adhesive. This allows more opportunity for differential movement between wall and tiles without loss of adhesion.

Round edge tiles, ventilators, soapholders, footrests and other accessories are available for use with wall tiles, but in a fairly limited range of colours.

Most Australian-made tiles are of plain colours but some local artists and overseas factories produce a magnificent range of patterns and part patterns which can be manipulated with site fixing to form larger patterns.

Mosaic tiles of square, octagonal and irregular shapes often incorporate gold-impregnated glazes to give special sparkle and richness. (Glass tiles are used for this decorative purpose also.) When laid, these tiles do not need to be perfectly flat and uniform, as variations in light reflections enhance their effect.

SUMMARY OF CERAMIC CHARACTERISTICS

1. Structural Strengths
The harder baked stonewares display good compressive and fair tensile stress characteristics, while products baked at lower temperatures are weaker in these qualities. Mostly, these items are not expected to support loads other than their own as wall facing materials, but floor tiles need to be strong in compression and able to resist abrasion.

2. Manufacture
Sources of manufacture and supply need to be checked before selecting ceramic products to ensure availability of selected samples.

3. Water
Unglazed earthenware products are usually absorbent, but they resist atmospheric chemical attack. Freeze/thaw effects need to be considered for unglazed ware in exposed locations. Glazed products should be non-absorbent, but defects in the glaze sometimes lead to water penetration and associated disfiguring problems.

4. Temperature Effects
Heat does not seriously affect ceramic products unless applied in a concentrated form to a small area. For large tiled areas, allowances must be made for expansion. The generally hard, smooth-surfaced materials tend to reflect heat, but the terra cotta type materials will absorb and store it, as do bricks.

5. Ultra-Violet Radiation
Most ceramic products and glazes are not seriously affected by ultra-violet light, but some slight fading may occur.

6. Electrolytic Effects
Ceramics generally are good electrical insulators and are often used for this purpose.

7. Acoustic Qualities
Because of the hard, smooth surfaces on many ceramic products, they tend to be very reflective to airborne sound. Spaces within tiled walls and floors thus become highly reverberant (as in many bathrooms) producing an echo effect; however, the density of these materials adds to the insulation of walls and floors against transmission of airborne noise.

Impact noise on tile floors can be transmitted through the bedding and concrete slab to spaces below.

Standards
AS 2049-1992	*Roof tiles*
As 2050-2002	*Installation of roof tiles*
AS 2358-1990	*Adhesives for ceramic tiles*
AS/NZS 3661	*Slip resistance of pedestrian surfaces*
AS 3958.1 & 2	*Ceramic tiles*

CSIRO Pamphlets
NSB 124	*Internal ceramic tiling*

METALS

Ferrous or non-ferrous; structural or non-structural; electrolytic action. Ferrous history and manufacture; terms; steels; jointing; steel products; stainless steel, references; summary. Copper and its alloys; brasses; bronzes; summary. Aluminium; alloys, characteristics. summary. Zinc; galvanising; zinc aluminium coatings. Lead.

INTRODUCTION

Most metals have to be mined and refined from their geologically based ores. The refining processes vary greatly and have developed the science of metallurgy.

For building purposes most metals are alloys, although the base metal is sometimes 99% or more of the mix. Alloys incorporate other minerals to produce certain desirable qualities not available in the parent metal alone. However, even small fractions of other minerals have significant effects on the properties of the parent metal so that almost all commercial metals are alloys.

The major base metals used in building work are iron, copper, lead, aluminium and zinc.

The metals using iron as their base are referred to as 'ferrous' and include the important building materials known as structural steel, mild steel, stainless steel, corrugated iron, sheet steel, etc.

The 'non-ferrous' metals include brass and bronze which are important alloys of the base metal copper.

Another way of categorising these materials could be as Structural and Non-Structural metals, using these terms in a very broad building sense. A brief summary on those lines is as follows:

Structural Metals	**Uses**
Cast Iron	Historical columns and frames from 1780 to late 19th century, included decorative work.
Wrought Iron	Pipes, decorative gates, grilles; major structural members during 19th century.
Mild Steel or Structural Steel	Rolled steel shapes, plates and members developed since 1890s in standardised forms for columns, beams, etc.
Steel Reinforcing Rods	In reinforcing concrete.
High Tensile Steels	Post-tensioning concrete; tensile members in structures.
Aluminium Shapes and Sections	Lightweight, temporary and permanent corrosion-resistant structures.

Non-Structural Metals	
Copper	In sheet and profiled forms for roofing, flashings, pipes for plumbing, electrical wiring.
Bronze	Special decorative hardware, grilles, doors, etc.
Brass	Plumbing, tubes, rods, hinges, hardware.
Zinc	Roof sheeting, flashings, die-cast hardware, protective coatings.
Lead	Roofing, flashings, old plumbing.
Aluminium	Tube, rod, extrusions for decorative and semi-structural purposes, window and door frames; folded sheet in various forms.
Stainless Steel	Rainwater goods, swimming pool fittings, urinals and sanitaryware, sinks, bench tops, etc.
Galvanised and Zincalumed Steel Sheet	Rainwater goods, sheeting, roofing. Fences
Cast Iron	Pipes, sanitaryware, decorative grilles.

Wrought iron, copper, bronze, brass, zinc and lead have been used in building work for centuries. The applications of steel and other metals in the last 150 years have caused major changes in the whole appearance and performance of buildings.

Centrepoint Tower, Sydney, exhibits the many roles of metals in 20th century building technology. High-tensile steel cables brace the rust-resistant structural steel core and floor cantilevers. The observation decks are sheathed in anodised aluminium & glass.

Electrolytic Action

This is an important consideration in using various metals adjacent to each other. Corrosion arises when moisture is present, causing an electrical potential difference between metals. Usually it is the metal at the negative (anodic) end of the galvanic series or scale which corrodes. The further the two metals are apart on the scale the more severe the corrosion is likely to be.

An expanded galvanic table to include common building alloys is as follows, beginning at the anodic (corroded) end:

zinc
aluminium
low carbon steels
alloy steel
cast iron
stainless steel
lead
tin
yellow brass
aluminium brass
red brass
copper
bronze
copper nickel

Separation of different metals is usually achieved by nonmetallic coatings or insertions.

FERROUS METALS - HISTORY

Iron has been in use for more than 3000 years in European and Asian civilisations. Steel produced by the introduction of a small percentage of carbon was probably an accidental development of the iron refining process. Primitive small-scale production methods have survived to the 20th century in parts of South-East Asia and India. Originally steel production was devoted mainly to swords and similar weapons.

Wrought Iron, hand forged on an anvil, was used widely for building purposes, especially in association with doors and windows, to provide hinges and protective locks, grilles, etc. which were a feature of important buildings in medieval times. Some uses of similar details still persist today, but very few really skilled anvil craftsmen remain. Hand-wrought work has been replaced largely by machine copies.

Traditional decorative wrought ironwork common before 1920.

Traditional Wrought Iron is 99.6% iron and 0.4% carbon. It is fibrous in structure and light grey of colour. It can be hammered, twisted or stretched when hot or cold. The more it is hammered the more brittle or hard it becomes, but it can be brought back to its original state by annealing (heating and then cooling slowly).

Cast Iron was introduced for structural purposes in England during the 19th century after the historic Coalbrookdale Bridge was built by Smeaton in 1776. Rolling mills for wrought iron were built in 1784. Industrial buildings up to seven storeys in height soon followed, using cast iron columns and wrought iron beams, with brick arched infills producing the floors and surrounding masonry walls.

A rolling mill for iron beams of I shape was built in France in 1845.

The famous Crystal Palace, built in London in 1851, was a classic application of cast iron construction technology to produce the largest-ever glass-house exhibition building in the record time of nine months.

Many large buildings constructed prior to World War I used cast iron columns.

Cast iron contains 2% to 4% carbon and is formed by pouring the molten metal into prepared sand moulds. When cool the sand is cleaned away and the iron remains but is very brittle and difficult to work. It is seldom used for structural purposes now because steel is more versatile and readily available.

Cast iron pipes, however, have continued to be used for water and sewerage because of their resistance to corrosion. Porcelain enamelled cast iron baths and basins are still popular. Rust does not form on cast iron as readily as on steel.

19th-century cast iron gateposts and finials with wrought iron gates and fence.

Cast iron verandah posts and lacework typical of late Victorian details.

Cast iron pipes here showing externally for waste water drainage are easily identified by the heavy collars at joints.

Commercial Steel Manufacture

Manufacture developed in Britain from 1856 to 1862, largely as a result of techniques introduced by Bessemer & Siemens. The first catalogues of standardised rolled steel sections and their structural properties were produced by Carnegie in the USA and Dorman Long in Britain about 1894, enabling engineers to predict the performance of beams and columns made of these components.

Light gauge sheet steel was produced by 1890 and quickly replaced the heavier gauge wrought iron sheets, particularly for galvanised corrugated sheets. The British colonies - especially Australia and India - were major markets for these products.

Mechanisation of processes in this century has increased the quantity, range and quality control of steel products particularly so that they are now used in every aspect of building work, frequently replacing more traditional materials, especially timbers.

Australian Steel Manufacture

The Hoskins family commenced manufacture in 1875 at Mittagong, NSW, then transferred to Port Kembla in 1927, and became Australian Iron & Steel. BHP opened steelworks at Newcastle in 1915, and in 1935 merged with AIS to become the dominant steel manufacturer in the Australian market.

John Lysaght was established in Bristol, 1857 and developed a strong local market for galvanised iron, and a central selling agency, then a factory in 1905 at Sydney which made corrugated iron, wire nails, barbed wire etc. so became a major customer for BHP Products. In the 1960's they developed the highly successful long life coating 'Zincalume' (45% Zinc & 55% Aluminium) for steel products which has largely replaced galvanising (100% zinc) for many factory made sheet metal products.

From amalgamations and mergers in the 1990's the Broken Hill Proprietary Co (BHP) became BHP Billiton, while Lysaghts and other subsidiaries became part of Bluescope Steel. Under this brand they lead the world in 'Zincalume' production and coating technology.

Australian steel for building purposes is also marketed in many forms by OneSteel, as rods, bars, pipes, wire, mesh, rolled steel sections, Litesteel beams, folded shapes, and stainless steel.

Carbon Content

The percentage of carbon in iron and steel affects greatly the working qualities and the production costs of the product.

Cast iron contains 2% to 4% by weight of carbon. It has high compressive strength but lacks the tensile strength needed for versatility in structural and building use. Cast iron products tend to be brittle.

As carbon content decreases, tensile strength and weldability improve, and the effects of rust in moist conditions increase.

Mild steel, commonly used for structural purposes, contains 0.15% to 0.25% carbon. It is reasonably cheap to produce, has good strength in compression, tension and shear, plus good weldability.

Wrought iron contains very low carbon content - 0.021% to 0.04% - and it is easy to work and weld, but it is rarely used for building today because of production costs.

SOME DEFINITIONS OF TERMS

Hardness is the ability of a material to resist indentation or abrasion.

Toughness is the property of a material which enables it to absorb energy at high stress without fracture.

Resilience or elastic toughness is the property which enables a material to absorb, without being permanently deformed, the energy produced by the impact of a suddenly applied load as in springs.

Malleability is that property which enables a material to undergo great change in shape without rupture under compressive stress as in hammering or rolling to produce special components.

Creep is the property which causes some materials under constant stress to deform slowly but progressively over a period of time.

Fatigue is a breakdown in the ability to resist normal stresses due to frequent overstressing or continuous stressing of the material.

Steel pipes, corrugated sheeting, gutters' and downpipes clearly evident in recent building additions.

58

STEEL

STEELS IN BUILDING

Structural steel, normally used in building work, is only one of the many steel alloys available from steel manufacturers. The minerals included with the base iron to produce the steel usually represent less than 4% but may consist of up to eight different elements in varying proportions.

Commonly incorporated chemicals include carbon, phosphorus, manganese, silicon, sulphur, nickel, chromium, copper. By varying the proportions of these elements the metallurgists can produce greatly varying qualities of steel.

Most steels are structurally useful because of their very high strength in tension, compression and shear. Like most of the metals, they are also very heavy, but their strength-to-weight ratio is high. They are also hard and tough. Mild steel usually has a dark blue-grey surface scale and is silver-grey and crystalline in the body of the metal.

Rolled Structural Sections

These sections are made by shaping a red-hot billet of steel into predetermined forms by passing it through a series of rollers. The common shapes available have been developed as a result of mathematical analysis of efficient structural forms and are now widely standardised into I, C and L forms and a range of sizes and weights.

The I sections originally produced had tapered flanges and curved edges, but these have been replaced largely by the universal beam sections which have a much more square-cornered profile. This makes for neater and easier connections, especially with welded joints.

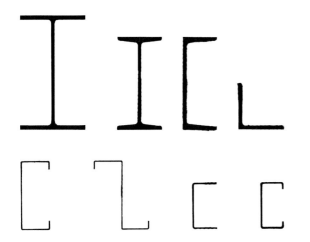

Some commonly available structural steel shapes; hot rolled (top) and cold rolled (lower).

The new LiteSteel beams developed by Onesteel and protected with zincalume.

Palmers Dogbone hollow flange beam (HFB) developed using continuous welding in 1994, available in high strength steel & depths from 200 to 450mm.

Rolled Steel Plates

Rolled steel plates are often used in conjunction with the I and L sections to make up large structural components, such as column bases, plate web girders, etc. by means of riveting, bolting or welding the pieces together.

High-Tensile Steels

These alloys are often used for rods and/or wire strands in prestressed and post-tensioned members where they need to resist the high stresses applied without the elongation which would be encountered with mild steel. Their composition often gives them characteristics and appearance very much akin to those of stainless steels.

Rust-Resistant Structural Steels

Specially rolled structural members can be ordered in rust resistant steel which includes a small percentage of copper. The Australian product is called 'Austen' steel and was used for some buildings for the steel manufacturer BHP. While this represents a significant saving in initial protection and periodic maintenance, it has not yet found widespread application. This steel takes on an overall rust-coloured appearance, but does not corrode beyond the initial oxidation.

JOINTING METHODS

Riveting

Until World War II, the commonest means of connecting structural steel pieces was by fixing red-hot rivets into pre-drilled holes and forming a head on the stem end by use of pneumatic hammers. Many railway bridges of the period demonstrate the technique. This method required steel angle pieces to connect plates at right angles to each other and was very cumbersome, slow and noisy compared with modern welding techniques.

Bolting

For on-site fixing of major components, steel bolts and nuts are often fitted into pre-drilled holes and tightened to a predetermined stress. Many structural members are assembled this way, such as column base plates to foundations, trusses to columns, etc.

Welding

The traditional blacksmith's system of fixing two pieces of similar metal together by hammering while red hot is very ancient and is still practised widely with wrought iron and steel. Modern welding for building purposes, however, developed from the oxyacetylene torch and the use of the electric carbon arc, both of which can produce extremely high temperatures sufficient to make adjacent surfaces of steel flow together and become one continuous unit.

A modern steel building frame employing many of the above members and including built up beams.

Welding allows joints in steel to be made at awkward angles as neatly as square junctions. In this example bolts are also used to connect cap and baseplates.

By welding shaped steel plates together, complex and structurally efficient forms can be neatly produced. The roof-supporting cables are high-tensile steel.

Riveted steel connections used in the Sydney Harbour Bridge were cumbersome and inefficient compared to modern welded joints. Note the huge (280 mm dia.) bolts which support the total bridge load and allow for movements due to expansion and contraction in the 536-metre arch.

Highly trained operatives are used to do this work and they can produce a great variety of weld types to suit many applications. As a result of technological developments and skills produced during World War II, welding moved out from the factory and into the post-war building boom as on-site work.

Welding is now the major method of jointing steel structural components. Joints involving light sheet steel can also be welded with special equipment but is usually a factory operation.

Connectors

A wide variety of specially formed connectors for use with standardised steel products on site connections is available. These include -
Screw-threaded connections for pipes.
Clamp-type connections for scaffolding and wire rope.
Bolts, nuts and washers.
Self-tapping screws and pop rivets for light gauge steel.

STEEL PRODUCTS

Steel Pipe and Tube

Circular steel pipes are available in a wide range of steels and sizes, thicknesses, etc., all made specially for building purposes. Steel pipes for water services have been used widely but need to be galvanised internally and externally to minimise corrosion. Even then these pipes have to be replaced at intervals of several years according to the chemical properties of the water supply.

Circular, rectangular and square steel tubes of various dimensions are commonly used for light structural purposes in roof members, furniture, etc. If protected from weather and well maintained, these members will give long service in many locations but are best not used in moist coastal areas unless protected by one of the zinc-based or plastic coatings.

Steel pipes used as structural components. Balustrades incorporate lightweight pipes and steel rods welded together.

Cold Formed Sections

Many products are made of sheet steel which is formed into shape when cold. By this means structurally useful L, C, Z and T sections for roof framing and similar purposes are made.

Sheet steel protected by an applied coating of Zincalume or galvanising is often used to give these light-weight members good life expectancy.

Sheet Steel Products

Most sheet steel products used in building are protected by galvanising, Zincalume or other high-performance protective coating so that the thin sheet does not rust away quickly.

These sheets can be formed into many shapes such as corrugated, ribbed, etc. giving them added structural strength and allowing for the construction of weatherproof laps where adjoining other sheets.

Modern continuous rolling mill processes allow these to be manufactured in great lengths, but usually they have to be cut down for transportation. The steels used in these products today are often much harder alloys than were used earlier, making onsite cutting and fitting more difficult than previously. Consequently suppliers usually measure carefully and deliver the goods pre-cut to minimise site work and wastage.

For the range and applications of sheet steel products and methods of jointing and fixing refer to Lysaght's Referee 1988 edition or its successor Bluescope Lysaght Referee.

Panels for complete building systems are now available consisting of a core of insulating material with faces of sheet steel, including stainless steel, with various epoxy or other plastic based protective surfaces. eg Robertson Protected Metal, & Kingspan Insulated Panels. Core materials can vary depending on the insulation performance required. Such panels are made to order to suit the building design.

Galvanised Iron

The well-known corrugated form of metal roofing, galvanised iron, derives its name from the early introduction of galvanised wrought iron sheets about 1840 before steel production became commercially viable. All galvanised sheet material produced in the original name of galvanised iron has persisted. (See also Galvanising, later, after Zinc).

Galvanized sheet steel is still produced, but for many uses subject to weathering Zincalume has proved superior and has superseded it. See Zinc Aluminium Coating later.

Steel Bars and Rods

Circular, square and rectangular steel bars or rods are used widely in building for structural and decorative purposes. Many balustrades, handrails, grilles and other items are fabricated from steel bars or rods, which have virtually replaced traditional wrought iron for these purposes. However, the steel is more prone to rust problems than is wrought iron. For most situations exposed to the weather these items need to be galvanised or well protected by a highly corrosion-resistant coating. As most of the joints are now welded these areas require special attention as the heating removes the natural mill scale on the surface which gives slight protection to the new metal' bars.

A century of changes in metal products come together at one point. Wrought steel, galvanised rods and hollow tubular sections in balustrades.

Reinforcing Steel

For reinforced concrete work, a wide range of specially milled steels is produced which includes deformed round bars, twisted square bars and electrically welded steel wire mesh. The main purposes of these items are to achieve maximum bond between the bars and the surrounding concrete and to minimise labour intensive bending of bars which was previous general practice. The cold working of some of these products also produces special tensile qualities in the steel which can be exploited in the structural design.

The steel wire meshes are particularly useful in normal domestic footings and floor slabs where loads are not great and the scale of the work does not warrant specialist reinforcement fixers.

A recent innovation in reinforcing rods produces a hardened and self-tempered surface layer around a ductile core of fine steel. These bars can be more easily bent and welded than others, and without loss of strength.

Light Steel Framing

Light steel framing using a range of studs, joists, trusses etc, has become widely used for domestic construction and office partitioning. The members are folded from zincalumed sheet, and often factory assembled into panel units for road transport to the site.

Light steel .stud framing as used in domestic frames and office partitions. Illustration by courtesy of BHP

STEEL

Straight, shaped and deformed steel reinforcement placed in preparation for a raft concrete foundation. Note also the ducts for post-tensioning cables used in this example.

Custom Formed Sections

By using cold forming, metal sections can be folded into shapes for specific requirements and in long lengths. Some of the shapes possible are illustrated on next page.

Duragal products are steel with a light galvanized coating which is not intended for long life in external building situations, but excellent for interiors and temporary work.

Zincsteel is flat sheet specially produced for internal components of furniture etc. needing consistent precision sheetmetal work.

Expanded Mesh

By partially cutting through sheet steel and then machine stretching the sheet, a variety of useful, sturdy and decorative forms of mesh are produced.

Wire Netting

Light gauge wire is woven into the familiar forms of birdwire, chicken wire, etc. which are usually galvanised and often used to support roof insulation, etc. in buildings. This material also forms the reinforcement for some concrete products such as ferrocemento.

Accessories

The steel industry has generally realised the need for a thorough development of systems for building rather than for 'components'. This approach has led to the supply of a wide range of accessories to simplify and minimise site work and gain advantage from the order of accuracy possible with factory production.

By this means steel systems have become very competitive with traditional on-site building techniques.

Care must be taken with fixings such as nuts and bolts to see they have corrosion resisting coatings equivalent to the material fixed. Electroplated zinc products look similar to galvanised articles but have much thinner coatings than galvanized protection on screws and bolts.

Protection of Steel

Most exposed steel, unless stainless, galvanized, zinc coated or Zincalume, needs to be protected by some effective means, usually painting. There are a great number of paint-on products available, offering varying degrees of protection, but at least one metal priming coat and two finishing coats are desirable.

Priming coats which contain an etching component are often recommended for the very smooth surfaces of zinc coated metals. Selecting the correct priming coat is particularly important. This base coat must adhere well to the steel, and be compatible with the finishing coats. Paint and chemical companies should be consulted and their product performance checked regarding unusual or critical situations. Choice of suitably protected metal or stainless steel fixings compatible with the main component is also an important factor in the life of any assembly.

Much of the Zincalume used in roofing and cladding is provided with factory finished coatings of paint baked on to the sheet. This gives a very extended life and attractive high quality appearance to the sheet material. This Bluescope product is called 'Colorbond'.

STAINLESS STEEL

Stainless steels are smoother, harder and more silvery than mild steel. They are a product of the complex metallurgical knowledge which grew with the steel industry in the early years of the 20th century. They are basically alloys of chromium and iron containing more than 10% chromium which have remarkable resistance to corrosion and heat. Other elements added to the iron-chromium combination produce special characteristics. The most important of these is nickel, which is used to produce 18-8 stainless steel generally used in building work. This alloy contains 18% chromium, 8% nickel, with small percentage proportions of carbon, manganese, phosphorus, sulphur and silicon added to the iron.

An early and prominent use of stainless steel was in New York in 1926 on the Chrysler Building, where much of the highly decorative and utilitarian features near the top of the tower and around entrance doorways used the new metal. Recent examinations of these details showed them to be in excellent condition despite more than 50 years of exposure in a very severe atmosphere.

Many forms of stainless steel are available and used in building, especially for pipes, angles and sheets which have been fabricated into many standard and specialised products.

One of the outstanding successes for the material has been for sanitaryware for public places - e.g. urinals, WCs, sinks, etc. where its ability to resist corrosion and abuse have been most apparent.

Practical and decorative uses are increasing, especially for shop fittings, where the high strength and good appearance - with minimal maintenance - are great assets. Various surface finishes are available.

For water supply, drainage, alcoholic containers and pipelines, in laboratories, kitchens, etc., it has proved to be an excellent material.

The structural strength, hardness and toughness of this material enable it to be used in very light sections so that the finished weight of articles is often remarkably light.

The outstanding structural advantage in the performance of stainless steel is in its retention of strength under heat and consequently fire. So far its costs inhibit wide use for general structural purposes, but the Centre Pompidou in Paris (1976) demonstrated some innovative uses in fire protection.

The range of standard components of stainless steel is increasing. Tubes of square, rectangular, circular, oval and other shapes have found wide application for building, furniture and industrial uses. The combination of strength and corrosion resistance is not matched by other products.

Site Work

Except for very simple drilling or cutting on site, all shaping, etc. must be done in suitably equipped factories. Leakproof joints can be made by welding, soldering and brazing, but site joints are usually brazed at approximately 700 deg C to 1000 deg C and special precautions taken with regard to cleaning and finishing. Various mechanical jointing methods are used where leakage is not a problem.

Stainless bolts, nuts and screws help to make for minimum maintenance installations.

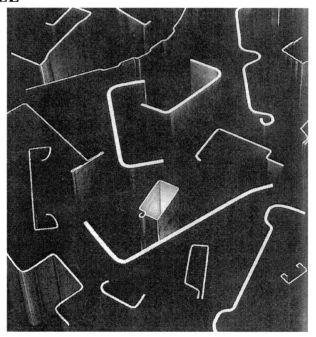

Some Custom Roll Forming shapes. Photo by courtesy BHP

Stainless steel tubular frames for large windows.

The complete destruction of mild steel roof trusses by fire is clearly apparent here.

An unusual precast concrete stair with stainless steel balustrade.

SUMMARY OF STEEL CHARACTERISTICS
Structural qualities
Structural steels generally have a density of 7900 kg/m3. Their structural qualities are very high: in tension 480 MPa; in compression 340 MPa. Compared with other traditional building materials this represents a tremendous combination of these important qualities for members to be used as beams and slender column's.

Stainless and high-tensile steels have higher structural strength, are harder and tougher.

Manufacture
Steel products usually have two or three separate factory processes involved before reaching the building site. Manufacturing is a large-scale operation which produces a great variety of standard shapes and sections which can be chosen to suit most building needs, and with minimal modification.

Consistently high levels of accuracy and quality are usually achievable with steel products. Structural components are fully prefabricated and only assembled and erected on site. Most components are made to measure in the factory.

Water Effects
Fully immersed steel does not rust appreciably. Water plus air is steel's worst enemy in building situations because it causes corrosion, commonly called 'rust', or ferrous oxide. Rust not only eats away the body of steel but also causes it to expand, thus cracking masonry and distorting adjacent surfaces. Rust also produces a typical reddish-brown stain that disfigures building surfaces.

Rust-resistant steels are available and sometimes used for structural components.

Stainless steels are not affected by water.

Temperature Effects
The coefficient of expansion of steel strain/ °C is 12. Normal variations can cause significant temperature movements in building components that must be considered in planning and detailing.

Steel's tensile and compressive strengths quickly diminish in temperatures above 500 deg C, so that steel used to support floors or other structural members must be protected from excessive heat such as is experienced during a fire.

Cold at or near freezing also changes some of steel's characteristics. This relates particularly to conditions under which welding might take place on metal near 0 deg C.

As steel is a good conductor of heat it can often be too hot or too cold for human comfort and needs to be protected, often by plastics, in many public locations where pain could result from direct contact, such as with handrails. Construction also creates a problem with condensation where there are differences between internal and external temperatures and sheet steel roofs are used. Thermal bridges via metals should be avoided between inside and outside temperatures wherever possible.

Ultra-Violet Radiation does not affect steel.

Electrolytic Effects.
Steel and iron are located centrally in the galvanic table of metals and this is beneficial in that extreme reactions do not normally occur with other metals. This relationship is important in the use of galvanising and other metallic protective coatings on steel. (See notes on Galvanising, after Zinc.) Nevertheless, some galvanic situations can cause serious problems with steel and specialist advice may be needed.

As steel is a conductor it must be insulated from electricity.

Acoustic Quality
Impact sounds are transmitted clearly through steel. Sheet steel provides very little barrier to airborne sound and may also contribute to noise levels by its potential to develop a drumming effect in thin sheets and by having a smooth reflective surface.

References
National Association of Steel Framed Housing (NASH)
www.nash.asn.au can provide advice re these products and systems.

Kingspan Insulated Panels Pty Ltd, St Marys NSW
www.kingspan.com.au

Stramit Building Products - www.stramit.com.au

SCHLENKER : Introduction to Materials Sciences; Wiley.
BLUESCOPE: The Bluescope Lysaght Referee.

'Steel Profile' is an excellent periodical produced by Bluescope Steel, which includes some useful technical information.
Numerous other pamphlets should also be available.

Standards:

AS 1250-1981	SAA Steel Structures Code
AS 1445-1986	Hot dipped zinc coated or aluminium/zinc, coated steel sheet
AS 1538-1988	Cold formed steel structures
AS 1554	Structural Steel Welding Code
AS 1562.1 -1992	Design & installation of metal roof & wall cladding
AS/NZS 1576: 1995	Scaffolding
AS 2180-1986	Metal rainwater goods- Selection & installation
AS/NZS 2179	Rainwater goods
AS/NZS 2312-1994	Protection of Steel from Corrosion.
AS 2327	SAA Composite Construction Code
AS 3623-1993	Domestic metal framing
AS 3703	Long span corrugated steel structures
AS 4040-1992	Methods of testing sheet roof & wall cladding
AS 4100-1998	Steel Structures
AS 4600-1996	Cold formed steel structures
SAA HB39-1992	Code of common practice for steel roofing
SAA HB48-1993	Steel Structures Design Handbook
SAA HB22-1992	Commentary on Structural Steel Welding Code
SAA HB31-1992	Handbook of building construction tolerances
SAA HB39-1992	Code of common practice for steel roofing

CSIRO Publications

NSB27	Arch bars and angle lintels for brick walls.
NSB72	Steel lintels for garage door openings.
NSB78	Some condensation problems.
NSB79	Corrosion of metals in building.
NSB80	Some common noise problems.
NSB82	Curtain walls.
NSB14 9	Cathodic protection.

Bluescope Steel Publications
www.bluescopesteel.com.au for available material
Hot Rolled & Structural Steel Products. 1994 Edition
Structural Steel in Housing First Edition Tables & Details
A guide to Cold Formed steel Purlins & Girts
Galvaspan Technical Bulletin for Purlin specifications
Bondek II Composite Slabs Manual
Unifloor Manual for suspended slab floors on sloping sites.
Custom Formed Sections
Colorbond Roofing and Fencing booklets
Colorbond Stainless & Colorbond XSE
The Steel Architectural Panel
New Zincseal;for interior furniture and components
Technical Bulletins particularly regarding roofing & walling products
Many technical pamphlets are available from local steel distributors covering building products and fixings.

Computer disks are also available for hot rolled and structural steel products; some files for CAD drafting etc. Contact Bluescope Steel Direct
Freecall 1800 022 999, www.bluescopesteel.com.au
Steel Profile Quarterly magazine aimed at architects, academics & structural engineers.
Steel Know Hour, Aimed at builders, plumbers, roofers & electricians.
Steel Hotline; Promoting improvements in manufacturing, applications; etc.

NON-FERROUS METALS

Copper

Copper was probably the first metal used deliberately by man, and this happened as early as 8000 BC. By 5000 BC its molten use in cast products had been established. Copper mines and refining were known to exist on the Sinai Peninsula about 3800 BC and beaten copper sheets, pipes and other objects used in buildings were produced about this time.

Copper is soft and malleable. It is readily worked into sheets, and on site, so its use in buildings has been widespread. As a waterproofing agent for roofs, etc., and as pipes for conveying water, it is still one of the most permanent materials available.

Copper's outstanding combination of useful properties led to its use in a vast range of commercial applications.

For general building purposes there is little difference among the corrosion-resistant properties of the various types of coppers, but their mechanical properties are affected considerably by small additions of other elements.

Copper is one of the purest metals produced commercially. Its assay generally exceeds 99.9%.

Apart from silver, which is slightly superior in conductivity, copper is the best conductor of heat and electricity of all known substances and is used widely in all areas of electrical engineering.

As copper is almost immune to atmospheric corrosion, the building industry uses large quantities of commercial copper for roofing, guttering, downpipes, flashings and dampcoursing. Fixings such as nails or screws for sheet copper must also be of copper to avoid electrolytic action. When left exposed to the atmosphere copper slowly takes on a pale green patina which is permanent and usually considered decorative. This can be accelerated by use of suitable chemicals.

Numerous factory-made articles used in building such as heating coils, hot water cylinders etc. aptly demonstrate the ductility and workability of commercial copper.

References

AS1432: 1973	*Copper tubes for water, gas, sanitation.*
AS1572: 1974	*Seamless copper and copper alloy tubes.*
BS1420: 1965	*Glossary of terms applicable to copper, zinc and other alloys.*

Copper ridge roll roofing, gutter, vents, etc., showing the permanence and versatility of this sheet metal in the hands of forgotten tradesmen

Old copper details and some modern ribbed roofing on a 19th-century building.

Various sized copper tubes as used in building services.

Bronze has long been a monumental and engineering material because of its hardness and resistance to corrosion. Steel has largely replaced it for engineering uses. Here steel engineering forms a backdrop for bronze sculptures.

COPPER ALLOYS

Bronze; an alloy of copper and tin, first appeared about 3700 BC in Egypt and its use soon spread over the Mediterranean area - to Crete by 3000 BC, to France and other parts of Europe by 2000 BC.

The Bronze Age of European history resulted from the production of alloys from these two base metals, which were used in various proportions from 16:1 to 8:1 to produce particular qualities in the bronze.

Some alloys produce metals as hard and tough as tool steel. In Renaissance times the art of bronze casting and sculpture reached new heights. The great bronze doors of the Florence Baptistery are one of the world's artistic masterpieces. Military and naval cannons were largely of bronze.

Bronze products are very durable and are still used widely in industry.

We are all familiar with coinage produced from a type of bronze containing 95% copper, 4% tin and 1% zinc. The colour of bronze varies from this coin colour to very dark brown depending on the alloy. Some bronzes exposed to atmosphere develop the distinctive green patina of copper roofs.

Brass was known and used in Roman times and has been developed also as an important utilitarian and artistic material by Asian civilisations.

Brass was usually alloyed from 7 parts copper and 3 parts zinc to produce a very attractive golden-coloured material which is softer than bronze and harder than copper and resistant to most atmospheric corrosion.

MODERN ALLOYS have increased greatly the range of secondary metals introduced to produce specific qualities for commercial production purposes. An outline of those currently available for building purposes follows.

The straight brasses are widely used because of their tensile strength and good fabricating characteristics. The addition of lead to such alloys increases machinability and produces the wellknown free-cutting alloys. The addition of manganese, iron and aluminium results in increased strength. The addition of tin and arsenic has a marked effect on corrosion resistance to sea water and similar environments.

Carefully controlled additions of manganese and silicon yield a family of forgeable bearing alloys with useful anti-friction properties for certain bearing applications. These alloys offer high strength and wear resistance in combination with good machining and forming properties.

Straight High-Zinc Brasses are the most extensively used of the copper-base alloys in strip and sheet forms. Alloys in this group are characterised by their excellent fabricating qualities and comparatively high tensile strength.

The straight brasses supplied locally contain between 30% and 40% zinc. A 70/30 brass is known universally as Cartridge Brass. Plumbers' hardware, automobile radiator tanks and refrigeration units are examples of its end use.

Other special alloys are suitable for cold-working operations such as blanking, stamping, forming and bending, and another variation is used for screws, nails, rivets, hooks and eyes.

Low Zinc Brasses
(Gilding Metals or Red Brass)

These metals contain from 5% to 20% zinc. They are characterised by their excellent cold working properties and beautiful colours, from coppery red to a rich golden colour. They are resistant to corrosion, possess good electrical and thermal conductivity, are highly ductile and may be hot-worked with ease.

Gilding metals are used in the form of strip and wire for the fabrication of decorative art metal and jewellery, or as sheet material are ideal for building facades. The selection of a particular alloy is generally governed by the required balance of strength, ductility and colour necessary for the specific application.

Leaded Brass

The combination of physical and mechanical properties suited to machined products is available in the leaded brasses. They are a group of 60/40-type alloys with the addition of small quantities of lead to improve machinability. A choice of extruded sections made of this material is available.

These alloys are available in many shapes and sizes. An attractive and relatively tarnish-resistant golden finish alloy is used extensively for door and window frames, curtain walling, mouldings, thresholds, grilles and interior decorative trim. Hinges and door locks are other typical applications.

Manganese Bronze

The manganese bronzes or high-tensile brasses are essentially 60/40 copper/zinc alloys to which other elements such as tin, manganese, iron, aluminium and lead are added to increase strength, improve resistance to corrosion and increase machinability. These alloys are suited particularly to marine hardware.

Phosphor Bronze

Phosphor bronze is a copper/tin alloy with a small addition of phosphorus which acts as a de-oxidant and improves the properties of the metal. This material is used mainly in engineering applications involving moving parts and high-wear resistance.

Silicon Bronze

Silicon bronzes contain copper, silicon and manganese. When alloyed with copper the metals markedly increase strength and endurance limits whilst retaining excellent weldability, ductility, durability and fabricating characteristics.

These bronzes can be produced with tensile strength comparable with that of mild steel, together with freedom from rust and the ability to resist corroding gases, liquids and vapours.

These properties make silicon bronze the preferred alloy for such applications as calorifiers, unfired pressure vessels, process vessels, welded tanks, and full mains pressure hot water systems.

SUMMARY OF COPPER METALS

Structural Qualities

The structural strength and hardness of the copper metals vary tremendously from the soft and comparatively weak pure copper to the very hard and extremely strong bronzes. The density of these metals is slightly greater than for steel - approximately 8500 kg/m3.

Manufacture and Availability

The major copper alloy production for building includes fixings (nails, screws), hardware (hinges, locks, doors and window pulls), plumbers' requisites (pipes and fittings), architectural sections (for doors, windows, etc.) and electrical goods (wires, cables, switchgear, etc.). Many of these items are available almost everywhere, whilst others have to be ordered ahead of need, especially with regard to architectural sections.

Temperature Effects

Copper metals in building situations have a higher expansion co-efficient than structural steel but considerably less than has aluminium (see Comparative Tabulations). The significant difference is between brickwork and the metals generally.

These alloys are good conductors of heat and lend themselves readily to heat treatment on site to create strong joints, as used in plumbing.

Water Effects

Copper and its alloys are among the most enduring materials known in the presence of water.

Water from copper pipes, roofs or gutters can cause green staining on adjacent surfaces if constant dripping occurs.

Research at Newcastle Hospital in New South Wales revealed that ultra-soft water had eroded copper pipes in that and other local buildings. A very rare defect for copper.

The malleability of sheet copper allows this type of embossed panelling to be readily produced. The mullions are aluminium.

The craftsmanship required to produce large bronze sculptures of this type has evolved over many centuries and is practised by a few experts. This group is mounted on a polished red granite base which is very water resistant.

ALUMINIUM

Electrolytic and Special Effects

Copper is the best commercial conductor of electricity available. It lies at the positive end of the galvanic scale and is therefore likely to be reactive with zinc, aluminium and steel in the presence of an electrolyte (water).

Isolation of copper from other metals to minimise electrolytic action can often be effected by the introduction of lead to separate the two metals. Electrical insulation requires non-metallic covers, usually rubber or plastics.

Atmospheric Effects

Ultra-violet light does not affect the copper alloys. Atmospheric conditions develop a patina of greenish or brown colour on these metals that can be accelerated if required by use of patented processes. No further protective coating is necessary.

Acoustic Qualities

Thin sheet roofing will transmit impact noise from rain. Where a solid sound or feel is required in association with these products a solid type or extruded section is desirable, but impact noise will be transmitted with all rigid metals.

Standards

AS 1589; AS 1628; AS 1645; AS 1718 & AS 4809; drainage. Each refers to copper and copper alloy plumbing products.

References

Austral Wright Metals is now Australia's leading distributor of stainless steel; aluminium; brass; bronze & nickel alloys etc, and should be consulted for details regarding these metals.

www.australwright.com.au

Aluminium is widely used for window mullions, frames and column cladding in this building facade.

ALUMINIUM

Aluminium is the most modern of the common metals, having been isolated in 1825 and introduced to the public at the Paris Exposition in 1885. The raw materials derive from very widely occurring clays and soils throughout the world, particularly the reddish-coloured bauxite.

Modern electrolytic methods of production were developed in 1856 in the USA and France. Mass production commenced in 1886.

Aluminium production usually requires vast quantities of electricity.

Popular uses came initially in kitchen utensils, where the reduced weight (approximately one-third of that of iron, brass or copper) was of significant benefit.

The aircraft industries are a major consumer of the strong, lightweight metal; and after World War II the metal entered the building industry as a large-scale competitor of traditional materials. It now enjoys a very wide range of uses, especially in glazed door and window frames, roofing and cladding materials, light structures, rainwater goods and decorative finishes.

The application of coloured finishes, some electrolytically bonded to the metal by the process called anodising, has greatly broadened its use.

Probably the major characteristic which has helped aluminium penetrate the conservative building market has been its suitability to production processes involving extrusion. By this means very complex forms for particular functions have been produced at economical cost.

When combined with the light weight and corrosion resistance of aluminium many such products have become economically advantageous, with performance superior to the traditional materials which they have replaced.

Aluminium is also a good electrical conductor, being up to 60% as effective as copper for this purpose, and is non-magnetic. It can easily be joined by bolting, riveting and welding and can be worked with faster cutting and drilling times than for steel. Welding is usually a factory operation under controlled conditions.

The very effective protective coating for steel called Zincalume consists of 55% Aluminium and 45% Zinc which is applied for sheet steel via a bath of the molten metal, creating a very strong bond to the surface of the steel.

Some aluminium extrusions showing the fine detail and complexity of shapes possible with this metal.

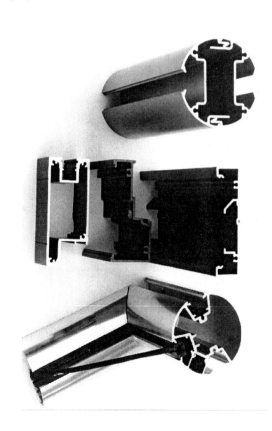

ALUMINIUM

Aluminium Alloys and Selection

Commercially pure aluminium has a comparatively low tensile strength of about 90 MPa, which limits its usefulness as a structural material. Large increases in strength can be obtained by alloying aluminium with small percentages of one or more elements such as manganese, silicon, copper, magnesium or zinc, with strength improvements being made by cold working or heat treatment. This is why alloys of aluminium are used for all structural purposes.

The Aluminium Development Council of Australia has adopted a system for designating wrought aluminium alloys based on four digits. The first digit of the designation serves to indicate the alloy group. The second digit indicates modifications of the original alloy or impurity limits. The last two digits identify the aluminium alloy or indicate the aluminium purity.

In general, the Aluminium Development Council of Australia alloys are similar to, if not the same as, Aluminium Association (USA) or British Standard alloys having the same number designation.

Alloys and alloy designation systems are covered to a greater depth in a book entitled STANDARDS AND DATA FOR WROUGHT PRODUCTS, published by the Aluminium Development Council.

The selection of the proper alloy for a specific application depends on the requirements of strength, durability and cost, the proposed fabrication method, and finally by the availability of the products.

Australian Standard AS1664 gives tables of minimum mechanical properties and maximum permissible stresses for a number of aluminium alloys of various tempers.

Designers are reminded that maximum permissible stresses are reduced where welding is involved.

The most commonly used aluminium alloy for structural and semi-structural extrusions is alloy 6063 in T5 or T6 temper. It is a medium-strength alloy generally favoured for architectural shapes because of its lower cost, good resistance to corrosion, excellent extrudability and responses to decorative and protective anodic surface treatments.

Where additional strength is required, alloy 6061 in T6 temper is recommended because of its high yield strength and good corrosion resistance. Alloy 6061 is marginally more costly than alloy 6063 but is permitted significantly higher maximum stresses by AS1664.

Series 5000 alloys are the normal choice for sheet and plate used in structural applications. Alloys 5052, 5251 and 5454 have sufficient strength for general structural purposes and have the benefit of excellent corrosion resistance. Alloys 5086 and 5083 are used predominantly in applications in combination with alloy 6061 extrusions, and where welding is the means of jointing.

As a general guide, alloy 5086 should be used in preference to alloy 5083 unless the additional strength of alloy 5083 in the nonwelded condition is required by the design.

For corrosion resistance, alloys 5052, 5251 and 5454 are marginally superior to alloys 5083 and 5086 in the hard tempers.

SUMMARY OF ALUMINIUM

Structural Characteristics

The high strength to weight ratio of aluminium is one of its notable characteristics, often producing building components or structures half the weight of equivalent components of steel. Its allowable tensile strength is 150 MPa (structural steel 160).

Manufacture and Availability

The world-wide manufacture of aluminium, the location of fabrication factories in key centres and light weight of components make this a readily available material in numerous forms. It is a very malleable metal.

There is a tremendous range of standard sheet, bar and extruded shapes available, some of which are restricted by licences. New extruded shapes are expensive because of the need for new dies, but for large projects this can become economical.

Water Effects

Rainwater generally only cleans aluminium, and it is resistant to normal atmospheric corrosion. Marine atmospheres and sea water can produce severe cases of galvanic corrosion and special advice should be sought for such locations.

Electrolytic and Special Effects

Aluminium is a good conductor of electricity. In the presence of moisture and contact with most other metals galvanic corrosion will occur. In many locations sufficient protection is given by application of pigmented zinc chromate paint to surfaces before contact is made.

Risk of contact with electric wiring must be carefully avoided.

Folded aluminium strips used as a facing, with extruded frames to doors and windows below.

Temperature Effects

Linear thermal expansion of aluminium is about twice that of structural steel (see Comparative Tabulations).

Structural strength diminishes at about 200 deg C, but cold and freezing conditions do not produce brittleness as they do in steel. Roof structures of aluminium do not usually require protection from fire, but because of expansion other critical structural elements may need protection.

Heat will flow through aluminium readily, but its surface feel to the touch is not as objectionable as steel's. Condensation on the inside surfaces of external components can be a problem. Thermal bridges between inside and outside should be avoided.

Ultra-Violet and Atmospheric Effects

'Natural' aluminium - i.e. in its normal silvery-white finish - will develop a thin, hard film of oxide when exposed to air. This film is highly resistant to weathering and the base metal remains protected against corrosion so that maintenance is minimised. Ultra-violet light does not affect this surface.

To give variations of colour and added protection, anodising processes have been developed for aluminium. These are available in black, white and various colours. Any of these applications need to be checked for weathering qualities, but many stable colours are now available that can resist U-V fading. White and light colours reflect heat, and dark colours absorb heat, with relevant consequences for expansion of components and insulating effects.

Acoustic Properties

Because of its light weight and metallic nature this material is not an effective acoustic insulator. Impact noise such as rain on an aluminium roof will be audible.

Coatings for Aluminium

Many aluminium components are factory finished by anodising or powder coating. This treatment protects the surface from the roughness which can develop on unprotected aluminium exposed to atmosphere. Items treated this way are often wrapped in plastic when delivered to the job, to protect them from being scratched or damaged by mortar etc. There is a wide range of colors available. Contact aluminium anodisers for information.

Further Information

The Aluminium Development Council of Australia is a major source for building products and suppliers information.

Standards

AS 1231-1.985	Anodized coatings for aluminium
AS 1562.1-1992	Sheet roof & wall cladding- Metal
AS 1664.1-2-1997	Aluminium Structures Design
AS 1765-1976	Aluminium Welding Code
AS 1903-1976	Reflective foil laminates
AS 2039	Methods, for testing anodic coatings on aluminium
AS 2048-1977	Code for aluminium windows in buildings
AS 2047-1977	Aluminium windows for buildings
AS 3500-2003	Plumbing & Drainage
AS/NZS 2179	Metal rainwater goods

CSIRO Publications

NSB 79	Corrosion of metals in building
NSB 82	Curtain walls
NSB 164	Roof' cladding
NSB 165	Cladding for small buildings

LEAD

Lead is probably one of the first metals man managed to extract from its base ore by smelting, as it melts at a very low temperature, 327.4°C, is soft and easily worked.

The Egyptians used lead coins and medallions and by Roman times the use of lead was widespread for utilitarian purposes. Lead water pipes have been found in Roman ruins, still in usable condition, testifying to the durability of the metal.

The term `Plumber' derives from these times; the Latin word for lead is `plumbum'.

The softness and ductility of lead enable it to be made into sheets and rolls, but its great density (almost 50% greater than mild steel) makes it heavy to handle and thin sheets and pipes will not even support their own weight. Because lead has a very high coefficient of linear expansion, fatigue and creep problems become visually apparent and have to be considered carefully in detailing.

Lead was historically popular for roof sheeting on the English medieval and later churches, and is still used extensively in England for roofing and roof plumbing. In Australia in recent years its use has diminished in building, but in some roof plumbing situations, subject to wind and rain, its malleability and weight make it still a most useful and recommended material.

Where roof water is to be collected for drinking, however, lead roofing should be avoided as it can produce lead poisoning over a period.

Lead is the most effective shield known to X-ray and atomic radiation and is used in buildings for these purposes, both in sheet form and as aggregate in concrete in atomic plants and laboratories.

For acoustic insulation in critical situations where the space required by other materials would be prohibitive, lead can also be very useful because of its high density and ductility.

Alloys containing lead are widely used as solders in plumbing. Many die-cast hardware items incorporate zinc and lead as the major metals. Lead can be readily cast into moulds and will reproduce minute details as in type metal used in printing processes.

A roof flashing detail for which heavy sheet lead is the ideal material to dress down over tiles.

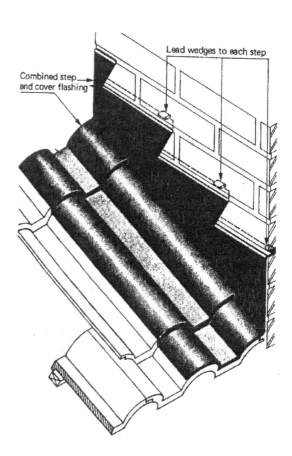

ZINC

Forms of Lead Available

Lead can be rolled to any desired thickness to create very thin sheets and foils, as well as the more common sheet forms in various widths of one to two mm thickness. It is usually supplied in rolled lengths. Pipe, rod, wire and other sections can be extruded in lead.

Laminations of lead combined with plywood, gypsum board or plastics are sometimes used for acoustic purposes, but need to be made to order.

Lead shot in spherical form is used as aggregate in concrete to resist radiation problems. This can be from 1 mm to 10 mm. dia.

Roof Plumbing in Lead

Modern lead usually is 99.9 % pure but some leads contain antimony which gives added stiffness and strength for a given thickness. The traditional quality of malleability and long life which gives lead its popularity for roof flashings, particularly over irregular surfaces such as terra cotta tiles, is dependent on the purity of lead.

Lead does not cause staining of adjacent surfaces and is generally more corrosion resistant than many other metals. It is one of the most durable of common metals. It must not, however, be used in conjunction with zinc aluminium-coated steel.

Molten lead can be used for the connections of cast iron pipes, where its softness allows for some flexibility at each joint and thus minimises leaks due to movements in the building or system.

Heavy-gauge lead, normally used in building, is available in various roll and sheet sizes and can be readily joined by seams to suit different situations. Holes or cracks can be repaired by soldering.

Thickness of sheet lead used to be defined in pounds per square foot, but with metrication the British Standard Code has given numerical gradings or BS Code numbers which roughly correspond with the old system. The following tabulation gives the comparisons for normal building applications:

BS Code No.	Thickness mm	Weight Kg/m2	Weight lb/sq ft
3	1.25	14.18	2.91
4	1.80	20.41	4.19
5	2.24	25.40	5.21
6	2.5	28.36	5.82
7	3.15	35.72	7.33
8	3.55	40.26	8.26

Movements in Roofing

In many roofing situations, lead is subject to wide variations in temperature due to direct sun exposure and night-time temperatures every 24 hours. The coefficient of linear expansion for lead is high at 0.0000297 per degree C and temperature ranges may often be 40°C. A 2-metre length of lead could therefore expand by more than 2 mm. Fixings to allow for this expansion to occur with minimal stresses are necessary to avoid fatigue failure, which causes cracking. Thickness of the sheet and sheet sizes are critical in this context. Consult references below for details.

Corrosion Resistance

Lead has high resistance to corrosion and can be used in contact with copper, zinc, iron and aluminium.

Some organic acids leached from timbers can be injurious to lead over long periods.

Lead sheet built into brickwork as damp coursing or flashings should be bitumen paint or plastic coated to prevent slow lime attack on the lead, which can cause pin holes and damp course failure after many years.

References

Lead Development Association (LDA) London, can provide information on lead products; and the ILA or International Lead Association handles items outside the UK.
  BS4513:: 1969   Lead bricks for radiation shielding.
  NZS 2116: 1966   Ingot lead for radiation shielding.

ZINC

Zinc in nature is usually in ores of limestone or dolomite, often associated with lead or copper sulphides. This accounts for the fact that brass, which contains zinc, has been produced since Roman times, although zinc as a pure metal was not isolated until about the 16th century. It was first produced commercially in England early in the 18th century and perhaps earlier in India and China.

A soft, greyish metal, zinc is ductile and can be rolled into sheets or thin foils which are flexible at atmospheric temperatures. Such sheets have been used for roofing and rainwater protection accessories. Zinc is also employed widely for electrical storage batteries.

By far the most important use of zinc in building is its application as protective coatings to steel products by way of galvanising, spraying, painting or electrolysis.

Zinc is also the major element in the alloys used for die casting by injecting molten metal into steel dies. The other metals in diecast products are usually aluminium 4% and magnesium 0.05%. Die casting is largely used for builders' hardware such as door furniture and for many parts in the automotive industry.

Zinc Roofing

Sheet zinc for roofs is not as permanent as is copper but has a life of approximately 50 years or more, depending on atmospheric conditions, gauge of metal used, etc.

The advantages of zinc are its malleability and consequent easy adaptation to complex forms. In Europe, especially in Paris, zinc roofs are a feature of the many awkwardly shaped pitched roofs where hand-shaped ridge rolls and welted seams are used. On rectilinear roofs, the modern factory ribbed sheets are more economical.

Because of the softness of zinc it is necessary to provide continuous support for this roof covering which tends to increase its cost, compared with the more popular steel systems. For some situations where replacement costs are high, zinc may still be a competitive material. Its fixing for awkward shapes requires an experienced and skilled roofing plumber.

The distinctive silvery crystalline pattern of galvanising on steel is clearly apparent in this illustration. The colours of zinc-aluminium coatings are similar but more silvery and finer in the pattern.

ZINC

GALVANISING

The protection of iron and steel from corrosion by coating with molten zinc. has been proved for a hundred and fifty years. This is called galvanizing. As the process requires the immersion of the product in a bath of molten zinc at approximately 500 deg C, the size of baths available has been a limiting factor for large building components. Recent increases in bath sizes available at major centres has greatly increased the potential for fully protecting large assemblies. This process known as 'batch galvanizing', produces a thick and very abrasion resistant skin which is strongly bonded to the steel, giving excellent weather protection. The coating penetrates to areas often unable to be painted

Sheet steel is also galvanized in the factory for further processing into many articles such as water tanks, electric light poles, ducting for air conditioning. Galvanized screws, bolts, nails etc. are also made for use with galvanized steel.

The mechanism by which zinc coatings protect steel has been explained as follows in the Australian journal ZINC TODAY. 'When a coating of zinc metal is applied to steel by the hot dip galvanising process a layer of almost pure zinc is metallurgically bonded to the steel surface. The steel is wholly enclosed in an envelope of zinc.

'Now the first protective attribute of zinc comes into play, its even corrosion rate in the atmosphere is slow and predictable, so that in a rural atmosphere, for example, galvanised steel will remain protected and unrusted for a hundred years or more.

'Engineers may therefore forecast their costs for maintenance; and perhaps in certain atmospheric conditions ignore maintenance altogether.

'That is the first, or barrier, protection for steel. The second protective service is what is called sacrificial protection: zinc being higher in the galvanic scale will sacrifice itself to protect metals below it in that scale. Should hot dip galvanised steel be accidentally damaged - and the zinc coating penetrated - the zinc surrounding that penetration will itself corrode to protect the exposed steel.

'In this context it should be noted that atmospheric moisture cannot penetrate the interface between zinc and steel. Rust cannot form beneath the zinc surface, as it does in a painted surface.

'The hot dip galvanising process has been serving industry for the protection of steel for over a hundred years, but it is only in recent years, here and overseas, that technology and zinc bath sizes have provided a service for major steel structures.'

Before specifying hot dip galvanising after fabrication it is desirable to check the limits of available galvanising bath sizes.

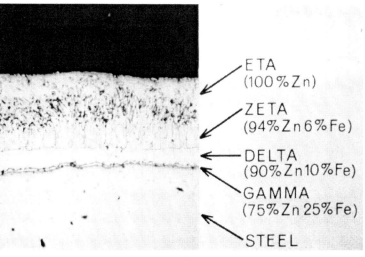

A microscopic photo/diagram of the metallic transmutation between steel and zinc created by hot dip galvanising.

For steel to be well protected against severe industrial, or marine environments within 300 metres of the sea, galvanising plus another surface coating is advisable. Sheet steel protection is best provided by galvanising or zinc aluminium coating followed by a PVC plastic coating applied in the factory. Any raw edges should be sealed on site.

The John Lysaght Research and Technology Centre has shown galvanised steel to be susceptible to rust from water drips containing salt, particularly when coming from a roof of dissimilar material, of either metallic or inert surface. See illustrations, and refer to Lysaght's Technical Bulletin SC 1-1 June 1983 for detailed recommendations.

A large fabricated component being immersed in a galvanising bath using a supporting cradle.

Other Zinc Coatings

Zinc is often applied to steel in other ways to provide protection adequate for less demanding situations than exterior weathering. One of these is termed Zincsteel by BHP and this is suited especially for furniture and cabinet manufacture. Zinc rich paints are also available for treating whole articles or small areas which may be damaged during the construction process.

Zinc Aluminium Coating.

Modern research, testing, & factory processes have produced a protective coating for steel combining zinc & aluminium called Zincalume (45% zinc & 55% aluminium) now applied to much of the sheet steel shaped into roofing, cladding, and light structural members for buildings. It is also a hot dip coating method, but the sheet steel passes through the bath at high speed in the sheet steel mill, and results in a thinner coating than with batch galvanizing. This is termed 'continuous galvanising'. Protection given to structural elements made from sheet steel treated by this process is not as effective as for batch galvanizing, but excellent for most normal conditions. Externally used 'C '&' Z' sections can corrode if exposed to marine or other corrosive conditions where salts can build up over a period. For any exterior structural steelwork exposed to weather near the sea, or in corrosive atmospheres, the type and thickness of the galvanized coating specified will determine the durability of the product.

The appearance of Zincalume is similar to that of galvanising but with a generally finer pattern. Some adjustment to plumbers' traditional techniques for jointing rainwater goods in galvanised steel are necessary to suit the modified alloy, but the products are now soundly established in the industry.

The very useful book 'Lysaght's Referee' should be referred to for technical information on these protected metal products. It includes methods of jointing, which differ from galvanized sheet, and include chemical adhesives. This book is now called Bluescope Lysaught Referee.

References

Industrial Galvanizers magazine 'Corrosion Management' May 1996 should be consulted for case studies.

Standards

AS 1397-2001	Hot dipped zinc or Zincalume sheet steel
AS 1562.1-1992	Sheet roof & wall cladding
AS/NZS 3750.9-1994	Organic Zinc primers
AS/NZS 3750.15-1998	Inorganic Zinc silicate paint
AS/NZS 2312-1994	Guide to protection of iron & steel
AS/NZS 4680-1999	Hot dipped galvanized coatings
AS 16274.4-1989	Abrasive blast cleaning

CSIRO Publications

NSB 79	Corrosion in metals
NSB 149	Cathodic protection

CONCRETE

Well-considered surface textures and finishes help to mask otherwise obvious variations and lifts of formwork with in situ concrete.

The base floors and columns are in situ concrete. Heavily moulded precast elements feature in the upper levels and encase the steel columns. Formwork, etc., for the lift shaft is also visible.

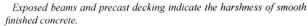

Exposed beams and precast decking indicate the harshness of smooth finished concrete.

CONCRETE

History, reinforced, quality control, terms and systems; references. Concrete products; blocks, tiles; kerbs; beams; pipes; fibre-reinforced products; references; concrete summary.

INTRODUCTION

More than 2000 years ago, the Romans developed concrete based on fine volcanic ash (pumicite) with hydrated lime cement, broken brick and stone, which they used in conjunction with their masonry to produce extremely strong walls and vaulted roofs.

The use of concrete largely died out with the fall of the Roman Empire and it was not revived until the development of commercial cements in the 19th century, following Aspdin's patent for Portland cement manufactured in Britain in 1824.

Modern concrete is an artificial stone produced in a plastic condition by mixing together 'aggregates' (sand and crushed rock); and a 'matrix' (Portland cement and water) all in controlled proportions. The chemical reaction between the cement and water produces strong binding qualities so that on completion of the curing time the resulting material is hard, dense and able to resist compressive stresses.

The characteristics of the concrete depend very largely on the qualities and proportions of the aggregates, cement and water. High compressive strength concretes usually include river gravels or igneous rock aggregates (basalt, granite, etc.) and only sufficient water to complete the chemical reaction.

Climatic conditions also affect the qualities of concrete. Protective measures have to be taken to guard against extremes of dryness, heat and cold, which can produce defects during drying out such as cracking, spalling or low compressive strength.

Concrete is usually ordered by specified compressive strength from ready mixed suppliers who calculate their ingredients to achieve that or better. Samples for laboratory testing are often taken at the site. Care is needed with delivery, placement and curing to ensure the specified strength is achieved.

Before ready mixed concrete was available, it was usual to specify concrete by volumetric proportions of the dry materials, e.g. 4:2:1 is four parts aggregate, two parts sand and one cement. The volume of wet concrete is considerably less than that of the dry coarse aggregate due to the filling of all voids by the compaction of mixing and placing.

In some conditions it is still convenient to site mix concrete, and dry mix in bags is a convenient way to obtain small quantities of concrete.

REINFORCED CONCRETE

History

The development of reinforced concrete (R.C.) commenced in the 19th century with some patents for the use of steel wire mesh in garden pots (1867), taken out by a Frenchman, Monier.

Many other innovators and experimenters have contributed to its practical application in buildings since then. Some notable people were Ransome (1885 onwards); Turner and Kahn, in the USA; Maillart, Perret, Le Corbusier and Nervi, in Europe; Young, Monash, Ove Arup and Seidler in Australia.

Reinforced concrete is now probably the most widely used building material in the developed countries of the world because of its adaptability and the ready availability of aggregates and steel reinforcements.

Reinforced concrete or ferro concrete is a combination of steel and concrete that is so designed it will utilise the compressive and fire-resisting capacity of concrete together with the tensile strength of steel.

Components that can resist tensile stresses resulting from bending, shear and torsion can thus be constructed effectively in R.C., due to the similar coefficients of thermal expansion and the bond developed between steel and cement.

Ferro Cemento is a special type of reinforced concrete, utilising a combination of fine wire mesh with cement and sand to produce comparatively thin members with a high proportion of steel, able to be formed readily into almost any shape. It was developed as a structural material by P. L. Nervi, in Italy, during World War 11 and subsequently used extensively by him in many buildings.

The material is used now in yachts more often than in buildings.

(See **Aesthetics and Technology in Building** by P. L. Nervi, for detailed descriptions of ferro cemento and its applications to major building projects.)

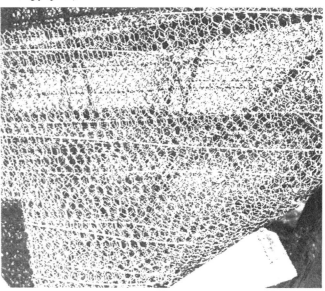

Galvanised wire mesh reinforcement ready to take the cement mortar to form a yacht keel of ferro cemento.

Reinforcements

Numerous high-tensile materials can be used to reinforce concrete, thus giving it good tensile as well as compressive strength if correctly placed in the members.

The most commonly used material has been mild steel rods, bars or mesh. To maximise the bond strength between the concrete and the steel, deformed bars are often used which give a physical keying effect as well as a chemical bond.

Steel wire mesh consists of light steel rods electrically welded together at each intersection so that there is no risk of misplacement or slippage of the reinforcement. This is very useful in concrete slabs for floors etc. The steel in mesh is frequently of higher tensile strength than normal mild steel.

Rods and bars are bent to engineers' drawings before being brought to the building site, and need to be clearly labelled for correct placement. Once in place, they are tied together with light steel wire at intersections. Clearances between the reinforcement and formwork must be carefully checked to produce the required concrete cover to the steel in the finished work to provide fire and corrosion protection.

In some situations reinforcing bars or mesh are galvanised to minimise the risks of rust developing. Little long-term observation of the effectiveness of this treatment on building elements is yet available. Some engineers doubt the benefits will be of long-range value; however, it has been the basis on which very fine wire meshes have been used to produce ferro-cemento yachts and other very thin skin structures for many years, which suggests it may be a developing technique.

Other non-continuous filament reinforcements not subject to rust are currently in building use and may transform some concrete reinforcement practices in the future. (See Glass Reinforced Concrete.)

Quality Controls and Specifications

The use of reinforced concrete in buildings is now controlled by building regulations and Standard Codes in most countries. Strict site supervision, however, is still necessary to ensure that these codes are observed and the design requirements are in fact achieved.

When ordering premixed concrete the supplier should be given the following information: strength, slump and cement type required; admixture permitted or required; maximum aggregate size; method of placement.

Corrosive environments can be found in the atmosphere, in water, in the ground, under traffic or industrial situations, and these can be very damaging to concrete even though complying with normal standards. When such conditions are found to exist expert advice should be sought. In some situations galvanised reinforcement can be used to minimise the serious effects of rust.

Density of the Matrix is the quality which determines porosity and consequently the likelihood of steel corrosion. With porous concrete the required code thickness of cover over reinforcement is still inadequate protection. Many buildings show this serious defect which is impossible to permanently repair.

For exterior concrete, especially columns, beams, cantilever slabs, etc. the specification and quality control need to be based on the essential concrete waterproofing and therefore density, more than compressive strength. This may involve testing of aggregates, cements available, and test cylinders to determine the necessary mix, followed by careful plant and job supervision.

It is extremely difficult to properly control site placement if more than one quality of concrete is involved in any one pour. After the initial set has taken place the concrete should be kept as damp as possible This particularly applies to floor slabs which have a large surface area exposed to sun & wind. They should be protected with a waterproof cover, or kept damp with a hose for several days. The ideal treatment is to keep water ponded on them for a week, but this is seldom practicable on a building site.

The working edge of an in situ pour showing reinforcement, concrete being screeded to finished level plus vibrators, etc., in use.

Waterproofing Additives.

Many additives are available which claim to make concrete waterproof. Some concrete batching plants will now supply concrete incorporating the correct quantities of selected materials. Before accepting the claims of manufacturers or suppliers careful investigations should be made to check how jobs using these materials have performed over the years. A paper 'Guarantee' is easy to give, but hard to enforce. This is most important for flat roofs which have a bad record for failures, which can lead to serious and expensive repairs, arbitration or legal actions.

These addditives usually derive from a calcium based chemical, and care must be taken to determine that the chemical will not react adversely with the steel or other component within the concrete to be treated.

A large and complex concrete pour in progress on a raft foundation. Delivery trucks, concrete pumps, pipelines and screeded concrete are evident together with raft and column reinforcement and ducts for post-tensioning cables.

CONCRETE

REINFORCED CONCRETE TERMS AND SYSTEMS
1. IN SITU

(a) Mass concrete (reinforced or unreinforced) usually in footings.

(b) Using steel reinforcement to resist the tensile and shear stresses where most advantageous.

In both types of application the shape of concrete is determined by the mould or form into which the plastic concrete is placed, this being its permanent location.

Forms can be - a trench in the ground;

- a temporary scaffold with timber, steel or other suitable platform, and walls that correspond to the height or thickness of member.

Forms should be leakproof and able to resist hydrostatic head, especially in columns and walls. Reinforcing steel is placed in the forms in accordance with design drawings and tied together with wire, and checked by the architect or engineer before the concrete is placed.

Concrete of the designed compressive strength is conveyed to the forms by means of hoists and barrows or concrete pumps and special pipeline. Soffit or horizontal forms should remain supporting the member for approximately 28 days. Vertical forms can be removed earlier. Concrete surfaces should be kept covered to promote slow, even curing, especially the top of slabs.

Common Applications

In situ reinforced concrete is now widely used in all elements of building structure and enclosure from the foundations to the roof, and especially for columns, beams, floor slabs, cantilever floors and balconies, stairs, etc.

Pavings and slabs on ground

In-situ concrete is widely used for roads, paths and floor slabs on the ground, usually from 100 mm thick upwards, depending on soil and load conditions. The surface can be finished in a variety of ways, such as rough (off the screed or broomed) trowelled smooth; or embossed with patterns from a mould. In roads the joints between poured sections are often shaped to key together. In large slabs allowance has to be made for shrinkage which will take place in drying out, to minimise the effects of shrinkage cracking. The concrete mix needs to be kept as dry as possible consistent with workability, as very fluid concrete shrinks more, becomes more porous and is weaker than 'dry' mix concrete. Waterproof membranes of sheet plastic are often used under floors to reduce the chance of moisture penetration from the ground.

Assembling timber formwork and reinforcement. Note the smooth and textured forms which will produce corresponding surfaces on the finished concrete.

Sawn timber texture imparted to finished concrete surface by chosen formwork material.

Joints in forms are often clearly visible in the concrete when stripped.

Defect in beam due to insufficient concrete cover over reinforcement causing steel to rust; sometimes referred to as 'concrete cancer'.

CONCRETE

2. PRE-CAST

Pre-cast components can be designed to serve in almost any building capacity, but they are most useful and economic in use where multiple repetition of forms is possible.

Use of steel, special concrete, plastic or composite forms is normal. Cost of forms is high, but multiple usage can make them economical.

Overall size and weight of components has an important influence on the transportation and lifting equipment needed at the building site, as well as on built-in hoisting eyes and reinforcement

3. PRE-STRESSED

'Pre-stressing' aims to place a tensile stress into the tension steel prior to the load being applied. This eliminates cracking due to shrinkage while drying, as the applied stress places the concrete in compression through the bond between steel and concrete.

The pre-stressing can cause 'camber' in beams, some of which is usually lost when the design load is fully applied. As a result the deflection normally encountered under load can be greatly eliminated, and the waterproofing and load-bearing qualities of concrete improved.

The system can be applied to both in situ and pre-cast work, but more commonly applies to pre-cast as special beds or anchorages are needed to enable the steel to be stressed prior to pouring concrete.

T-beams for floor slabs are a common application of this system.

Historical note

The development of pre-stressed and post-tensioned structural components is due largely to the pioneering work of Edouard Freyssinet, a French engineer who persevered with his ideas from the early 20th century until the practicality of his innovation was clearly demonstrated in the reconstruction of bridges, etc., in Europe after World War II. Freyssinet's name is closely associated with some of the techniques and equipment used in stressing the wire or cables used.

POST TENSIONING

Post tensioning can be applied to individual members or to a number of members by casting ducts into the concrete designed to allow the later insertion and tensioning of cables. When the required concrete set has developed and tension cables threaded and stressed, cement grout is forced into the duct so that it sets and fixes the tension into the cables and hence the member(s) into compression.

This technique uses the member(s) as the compression bed and hence can be used on site more readily than pre-stressing. In this way multiple pre-cast members can be brought together to act as one member. It is especially useful for beams and components of large dimensions or long spans.

This technique has been used in local buildings.
- Sydney Opera House (roof members).
- Trade offices, Canberra (main and secondary beams).
- Gladesville Bridge 300-metre arch span.
- New South Wales Institute of Technology Tower, Broadway, Sydney (two-way beam and cantilever system).

Techniques for demolishing such buildings may involve very great problems for future generations.

SLIP FORMING

This term refers to the process of placing in situ concrete into a continuously moving set of forms, eliminating the need for breaks in the concrete placement to allow for the removal and relocation of similar forms in 'lifts' of, say, 2m to 4m in height, as is normally the case.

Originally developed for the construction of wheat silos and similar continuous vertical containers. In the 1960's the system was adapted for use on lift shafts and stair wells, etc. in high-rise buildings, allowing for doorways and other holes to be included in the concrete walls. Numerous office building lift cores have been constructed by this technique. New South Wales State Office Block was probably the first in Australia about 1964.

An in situ base structure supporting upper elements of precast concrete in which the joints are clearly defined.

Most major components in this building were precast. The circular discs conceal post-tensioning cable ends, as this technique was widely used in the building to achieve large clear spans.

Reinforcement prepared in precasting yard. Note post-tensioning ducts.

Very careful control of ingredients, formwork, placing and finishing are necessary to produce the quality of in situ concrete achieved in the High Court in Canberra.

Preparing cables for tensioning.

Post-tensioned cables extend beyond concrete prior to trimming.

CONCRETE

Complex and decorative ribbed ceiling achieved by use of ferro cemento moulds as permanent formwork.

Heavily textured in situ concrete columns contrasting with smooth `C-shaped precast beams.

Bush hammered concrete surface and rough concrete block provide contrasting internal finishes.

Smooth concrete surface being removed by bush hammering.

Method of achieving heavy texture by hand hammering finely ribbed concrete.

A finely modelled and textured facade made possible by repetitive precasting.

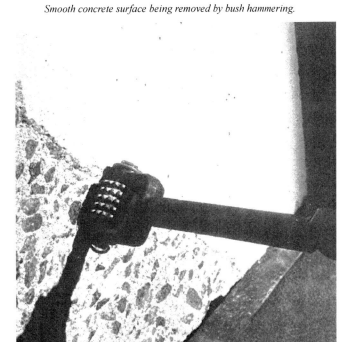

CONCRETE

TILT UP

Tilt up construction consists of a concrete floor slab used as the principal casting bed and site cast wall panels used as loadbearing elements capable of supporting the roof construction. These are usually cast full height, with no horizontal joints, on the floor slab then tilted to their vertical position when cured.

It is advisable to maintain simplicity of finish and single or 2 storey construction. Cost and construction time are dependent on good job planning by the builder. Economy is dependent mainly on intelligent use of cranes for relocating slabs from horizontal to vertical positions and no more than three storey construction.

Wherever possible, the panels should be used as loadbearing elements supported on simple pad or strip footings and tied into the floors by protruding reinforcement.

See Constructional Review, August 1980 (Cement and Concrete Association of Australia) for examples.

LIFT SLAB

This is a method of producing a multi-storey building of repetitive floor plans by pouring all of the floor slabs on top of each other on the ground, then jacking them up by using the in situ columns as hoisting masts.

The method was popular during the 1960s.

Other methods that developed superseded lift slabs, and it is not often heard of now. The theory was that major structural floors could be built without formwork, and the use of cranes or hoists was minimised.

LIGHTWEIGHT CONCRETE

Where compressive strength is not a critical factor and the total dead load needs to be minimised, lightweight concrete is sometimes used. This can be produced by using lightweight aggregates such as vermiculite (an expanded shale product), or by aerating the concrete chemically to introduce tiny bubbles throughout the matrix.

Technical References

Library shelves abound in volumes written about concrete. Cement and Concrete Associations produce excellently illustrated information and periodicals with practical articles for the building industry. Standards Associations and Local Authorities have developed numerous codes relating to concrete in building construction.

Relevant Australian Standards are as follows:

Standards:

AS 1012	Methods for testing concrete.
AS 1302-1304 1991	Steel reinforcing bars wire & mesh for concrete
AS 1310-1314	Prestressing materials.
AS 1379:1973	Ready Mixed Concrete.
AS 1465:1974	Dense natural aggregates for concrete.
AS 1467:1975	Lightweight aggregates for structural concrete.
AS 1478 & 1479 1973	Chemical admixtures for use in concrete.
AS 1480:1982	Rules for the use of reinforced concrete in structures.
AS 1481:1978	Rules, for the use of prestressed concrete in structures.
AS 1509:1974	Rules for the design and construction of formwork.
AS 1510	Code of practice for control of concrete surfaces. Formwork
AS 2327	SAA Composite Construction Code
AS 2876-1987	Concrete kerbs & gutters
AS 3600-1984	Concrete structures. Supplement available
AS 3610-1995	Formwork, for concrete
AS 3735-1991	Concrete structures for retaining liquids
AS 3972-1991	Portland & blended cements
AS 2870-1996	Residential slabs & footings
AS 3850-2003	Tilt up pre-cast elements for use in buildings
ASA 175-1970	Concrete kerbs & gutters
SAA HB 64-1994	Guide to Concrete Construction
SAA HB 71-1995	Concrete Design Handbook
SAA MP 20	Admixtures for concrete
HB 67-1995	Concrete practice on building sites
PP 58A	Concrete professional package

CSIRO Publications currently available regarding concrete NSBs

9	Pier & beam footings
28	Good concrete practice
36	Domestic floors
40	Suspended concrete
65	Shell roofs
92	Lintels for garages
100A & B	Admixtures
102	Surface finishes
105	Mixes
108	Off the form finishes
112	Suspended Concrete slabs
115	Curing compounds
125	Prestressed concrete
133	Aggregates for concrete
140	Concrete paths &paving
144	Pouring breaks
157	Placing reinforcement
162	Coral as an aggregate
174	Concrete paving
175	Toppings for concrete floors

Cement and Concrete Association of Australia
Concrete Information Publications & price list.

A concrete block plant showing stacks of various aggregates and rotating dragline.

Palleted stacks of blocks ready for delivery.

CONCRETE PRODUCTS

Due to the ready availability of cement, aggregates and reinforcements needed for concrete products, factories have produced a wide range for building purposes. These products often have the backing of large manufacturers with good quality control, testing facilities, continuous research and development. Their standards are generally good, and the products often better than can be produced by on-site labour and conditions. Control of curing is an important consideration which can often be carried out much better in a factory than on a building site.

For some products such as standard concrete blocks steam curing is often used to overcome quality variations which could develop with atmospheric curing.

Some of these products are in direct competition with more traditional building materials. Others have introduced new techniques and concepts which have become widely adopted. Consequently it is often necessary to compare the desirability of in situ construction with pre-cast concrete products. Both can be combined readily into the construction process.

Concrete Blocks

Concrete blocks for building are now available in a great variety of shapes and sizes and in both solid and hollow forms. Structural concrete blocks for masonry walling have been developed on a modular basis using 100 or 200 mm as the wall thickness, 200 mm as the unit height, and 400 mm as the unit length. These are now very widely used throughout the world. These blocks are usually hollow and unreinforced, but they are so designed that either vertical or horizontal reinforcement can be incorporated into the built wall to provide bending strength over openings, or to resist wind or other damaging stresses likely to crack or displace normal masonry. They are thus a very versatile masonry-type unit which allows the building designer tremendous scope to produce complete buildings with a minimum variety of materials.

In use, these blocks shrink and swell as a result of heat and moisture variations so that allowances have to be made for this - in external work particularly. Manufacturers' literature usually explains this, and shows some effective details for construction. One advantage these blocks seem to have over bricks is their ability to resist freeze/thaw effects in cold climates.

The natural colour of these blocks tends to be grey, and their texture varies slightly with aggregates used, but manufacturers now produce many coloured, split and ribbed blocks able to make a variety of textures for surface finishes. Many fine buildings have now been designed and built which demonstrate structural and visually effective uses of these 20th century materials.

Special Purpose Blocks

There are many special purpose blocks available in a variety of sizes and patterns for decorative and ventilation grilles. Some small manufacturers will make specials to the designer's requirements, provided the cost of the moulds is covered in the first order.

Paving Blocks

Concrete paving blocks are now available in great variety. The introduction of interlocking systems able to be lifted and readily relaid has greatly increased the attractiveness of these types of pavements for both vehicular and pedestrian areas.

Suitability of the product for the anticipated traffic must obviously be a prime consideration, and major manufacturers usually have test results available for their products, together with installation specifications for differing requirements.

Roofing Tiles

Concrete roofing tiles based on the Marseilles pattern have now become almost as commonly used as terra cotta tiles and are available in a wider range of colours and profiles. While these tiles are generally heavier than the terra cotta products, for most domestic purposes this is not a critical factor. Some absorption and colour problems which plagued their introduction into Australia in the 1950s have been overcome and the modern product is widely accepted as an alternative to terra cotta tiles. In locations such as ocean frontages they are not affected by salt spray.

A display of various patterns available in concrete roofing tiles.

Contrasting textures of in situ concrete and rock faced block walling are a feature of this building.

CONCRETE PRODUCTS

Interlocking paving blocks being laid. Many patterns are available.

Autoclaved Aerated Concrete (AAC) Blocks are made from chemically produced and steam cured lightweight concrete They are of a near white colour, easy to handle, accurate to shape and size, and can be cut and shaped with normal hand tools. Their fire resistance, thermal and acoustic insulating performance can be a major benefit in many situations. They are laid up using special adhesives in contrast to more conventional masonry's thick mortar joints. Their major disadvantage is their softness, which allows them to be readily damaged by vandalism or accident. By cement rendering they can be made reasonably resistant to normal abuse.

Kerbs and Gutters

Pre-cast kerbs and gutters have become an important item in site works and landscaping. Various shapes and sizes are available in pre-cast units for roads, industrial or domestic situations.

Steps and Stairs

Numerous pre-cast steps and stairs are available that can greatly simplify site work for these labour-intensive items. In some high-rise buildings stair flights have been completely pre-cast and hoisted into position, enabling them to be used immediately for access as work progresses.

Tee Beams

Pre-cast tee beams exploit the adaptability of concrete for expressing efficient structural shapes. They are often pre-stressed and available for various conditions from standardised forms. Their use minimises on-site formwork for floors and provides an immediately usable deck when placed in position. Topping slabs and finishes are usually applied later.

Ferro Cemento Moulds

Using adaptations of the techniques developed as ferro cemento by Nervi in Italy, many buildings have ferro cemento as permanent formwork into which structural reinforced concrete is placed. Some such moulds are specially designed for the job while others can be adapted from standard sized forms.

Concrete Pipes

The Australian company Humes Ltd. pioneered the manufacture of spun reinforced concrete pipes and has since expanded into an international organisation. These and similar pipes are available in a wide range of sizes and are used extensively in civil engineering projects and large-scale building works, especially for rainwater drainage, service ducts, etc.

Retaining Walls

Specially designed interlocking elements are available to construct retaining walls suited for many landscaping and site work situations. These enable such walls to be constructed without the need for skilled labour, formwork, etc., can be vibration and pressure resistant, and tolerate minor earth movements. Some patterns allow for curved and raked back walls. Consult local product manufacturers for illustrated technical literature.

Pre-cast products generally

Many other pre-cast concrete items are used in buildings, but the foregoing outline gives some idea of the range, and this is changing constantly and, of course, varies in different locations.

Special care is needed to protect surface finishes and corners of pre-cast concrete items during transport and handling, erection - and thereafter.

With modern transport, however, many items are produced far from their ultimate destination. The economics of handling at the building site often becomes an important consideration, especially if the project is a small one without the need for heavy lifting equipment. Special provision for lifting heavy units often has to be incorporated into the original reinforcement design.

National Precast Concrete Association of Australia produces a pamphlet 'National Precaster' which provides useful information on technical details and current projects.

High retaining wall using interlocking blocks which allow for minor earth movements and decorative landscaping.

Hollow concrete block walling which can be reinforced horizontally or vertically for structural purposes as in retaining walls.

CONCRETE: THE SUMMARY
Structural strength

The compressive strength of concrete can be designed within limits from 14 to 50 MPa to suit the particular situation. The tensile strength is a product of the location and type of the reinforcement, together with the shape of the structural member relative to the forces acting on it and the bond between cement and reinforcement.

Compressive strength and life of a member can be affected seriously by cracked and porous concrete, or inadequate cover to reinforcement.

Concrete cover over reinforcement must be sufficient to eliminate water penetration corroding the steel, and resulting expansion leading to concrete cover being 'blown off' in what is called 'concrete cancer'.

Manufacture

Both in situ and pre-cast methods of manufacture are used widely. In both procedures careful attention to design details, location of designed reinforcement, placement and consolidation of concrete and subsequent curing are important considerations affecting performance and durability.

Water and freeze/thaw effects

The water/cement ratio is critical regarding the resultant strength of any concrete product. Generally, only sufficient water for workability should be used.

Good, dense concrete is very water resistant but the real danger of water penetration occurs with shrinkage cracks which develop during curing. Careful protection of fresh concrete is essential to minimise cracking. Density of concrete should be a major criterion for all concrete exposed to weather.

Most concrete products are reasonably resistant to freeze/thaw effects after initial drying. During drying protection is necessary.

Temperature effects

Concrete and steel have roughly similar coefficients of expansion, therefore their structural bond remains despite temperature variations.

Absorption of heat by massive concrete is slow, but the heat is stored and slowly emitted after the ambient temperature drops. Expansion joints are necessary in most large structures. Concrete develops high fire-resisting qualities so that it protects encased steel reinforcement from loss of strength due to exposure to fire temperatures. Required concrete cover for various structural members to develop fire-resistance rating is usually defined by Standard Codes and building laws.

Ultra-violet radiation

This has little effect on concrete products except to produce a slight whitening of the original grey colour. This is frequently offset by atmospheric discolouration.

Electrolytic or other special effects

Concrete is normally an electrical insulator when dry.

Water penetration over long periods through cracks can lead to white crystalline build-up of dissolved lime from the cement. Atmospheric pollution usually shows up clearly on smooth concrete surfaces. Many textured surfaces have been developed to offset this undesirable quality.

Concrete floors in industrial situations can be seriously affected by material spillage and traffic wear so special finishes or toppings are sometimes needed.

Acoustic qualities

Being a reasonably solid heavy material, most concrete floors and walls reflect airborne sound. Low frequency and impact sound, however, can be transmitted through concrete unless a positive break occurs in the members.

As an insulator against airborne sound, concrete is comparable with other masonry materials of similar density.

Smooth concrete surfaces tend to cause high reflectivity of sound, and a long reverberation time for enclosed spaces.

References
Concrete Block Design Manual - Cement & Concrete Ass'n
Masonry Code of Practice -Association of Consulting Structural Engineers.

Standard Codes
AS 1475-1983 SAA Blockwork Code
AS1500-1974 Concrete building blocks.
AS 1653 Calcium silicate units
AS1757-1975 Concrete interlocking roofing tiles (without weathering check)
AS1759-1975 Concrete interlocking roofing tiles (with weathering check)
AS 2700 Masonry Code
AS 2733-1984 Concrete masonry units

CSIRO Publications
BTF 08 Concrete Mixes & quality control
BTF 09 Design & Control of concrete mixes
NSB 128 Bricks & Blocks
NSB 174 Concrete paving

Technical Studies
No. 3 Cracking in brick and block masonry
No. 85 Hollow concrete blocks.

Part section of a building incorporating in situ, precast, tilt slab and prestressed components.

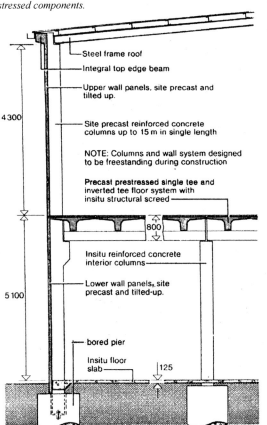

Steel frame roof
Integral top edge beam
Upper wall panels, site precast and tilted up.
Site precast reinforced concrete columns up to 15 m in single length
NOTE: Columns and wall system designed to be freestanding during construction
Precast prestressed single tee and inverted tee floor system with insitu structural screed
Insitu reinforced concrete interior columns
Lower wall panels, site precast and tilted-up.
bored pier
Insitu floor slab
4 300
800
5 100
125

Some widely available precast beam, slab and walling forms.

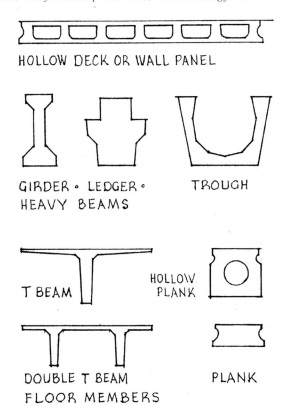

HOLLOW DECK OR WALL PANEL

GIRDER · LEDGER · HEAVY BEAMS

TROUGH

T BEAM

HOLLOW PLANK

DOUBLE T BEAM
FLOOR MEMBERS

PLANK

GLASS

History; building use; glass production; raw materials and manufacture; window glasses;
laminated; other specials; care and cleaning; properties; summary.
Glass products; fibres; textiles, foams; GRC; porcelain enamel; electrical goods; bricks.

INTRODUCTION
History

The glass industry is very ancient with records dating back to the Egyptians more than 3500 years ago. It is known that at least 6000 years ago - long before he had learned to smelt iron - man knew how to make glass.

The first glass furnaces were volcanoes, and the first man to use glass took it from the ground already fused and cooled. Natural glass formed by volcanic action is called obsidian and can be found in many parts of the world. It is usually black and translucent and was probably used to make arrow and spear heads, knives and razors.

The first man-made glass was in the form of a glaze - a mixture of sand and minerals heated and fused onto the surface of stone or ceramic objects in an oven, thus producing a hard, shiny outer layer.

By 1500 BC, man was forming glass beads and jewels and making small containers by dipping a sand core into molten glass. Glass threads applied while the glass was still hot gave the containers striking beauty.

Glass made before 100 BC was seldom transparent and it usually contained impurities and imperfections. However, the qualities of transparency and freedom from colour were unimportant for the manufacture of inlays, imitation stones and cored vessels. Shapes and uses became more varied with the introduction of the glass blower's pipe, which came into use about 50 BC.

Until modern times there has been little basic. change in the constituents which are fused to make glass. Roman glass manufactured during the time of Christ contains almost the same ingredients and proportions of ingredients as soda-lime glass used in today's bottles.

Until the 17th century, the only real advance was in the selection and purification of these ingredients. The pipe remained in use exclusively for producing all blown glass up to the 19th century and is still used for many types of modern glass working.

The art of decorating glass reached a high degree of perfection between the 2nd and 3rd centuries, but it declined between the 5th and 11th centuries. However, the use of stained glass for church windows (which commenced in the 10th century) is well known, and fortunately much of it is still available for public appreciation.

At the time of the Renaissance in Italy, glass making was considered a most beautiful and skilful art. The glassmakers were highly regarded socially. The manufacture of glass was considered to be a great secret; anyone divulging its composition was severely punished.

The island of Murano, near Venice, was the carefully guarded centre for the famed Venetian glass during the period AD 1250 to 1650. It was here that clear, colourless glass was again developed after being unavailable for centuries. Because of its resemblance to natural crystal it was called 'crystallo'. It was used for vessels, spectacle lenses, ships' lanterns, windows and mirrors.

By the 17th century the secrets of manufacture had spread to other parts of Europe and innovations in manufacture led to sheet glasses being evolved that were more suited to building applications.

In Elizabethan England a tax was placed on windows containing glass.

Glass for Building

The size of sheet glass was very restricted because the techniques of producing flat glass developed from blowing a cylindrical form, removing the ends and flattening out the curve to make the sheet. Alternatively, flat glass could be spun from a viscous globe to form a flat circular sheet with a central bull's-eye.

Each of these methods produced glass containing flaws or bubbles, but they were in use until recent times.

Cast and rolled plate glass was first developed in France in 1688 to 1702 and the influence of the development is clearly seen in 18th-century French architecture with its lavish use of glazed doors, windows and mirrors.

Use of flat sheet glass in botanical glasshouses led directly to Joseph Paxton's design for the famed Crystal Palace in 1851, which demonstrated the vast potential of glass as an architectural material. At this time the largest sheets available in England were 1200 min (4 ft) long, and the whole project was designed around this sized modular unit.

In the early years of the 20th century improved communications between the Belgian, British and American manufacturers of glass led to rapid improvements in techniques. Sheet glass was produced by drawing directly from the bath of molten glass, and various techniques were used to overcome some of the defects and difficulties associated with the method.

Improvements in the 1920s produced continuous plate glass which still required grinding and polishing to achieve flat, parallel faces free of distortion and up to 2400 mm (8 ft) wide. A later innovation in 1938 produced a continuous ribbon process in which the sheet was simultaneously ground on both sides then polished.

By 1959, Pilkington Bros. in England developed the Float Glass Process, where the ribbon of glass is floated over a bath of molten tin. This is now the most popular and efficient process available and produces glass of high quality and eliminates the grinding processes previously needed to produce distortion-free glass sheets.

Australian Glass Industry

Commercial glass manufacture did not really commence in Australia until 1866, when it began in both Sydney and Melbourne.

In 1916 Australian Glass Manufacturers Co. Ltd. was established by a consolidation of several small Sydney, Melbourne and Adelaide firms. By the 1930s a great variety of glassware was manufactured locally, including much that was needed for the building industry.

In the 1950s AGM developed into Australian Consolidated Industries, which in 1972 merged with the English firm of Pilkington to establish a float glass factory at Dandenong, in Victoria, which commenced production in 1974.

Pilkington ACI was the only local manufacturer of float glass and several other types used in the Australian building industry until 2008. CSR acquired Pilkington and DMS Glass in 2007, and this manufacturing merger produced 'Viridian Glass' as the new major player. Australia now produces most of the types and quantities of glass required, and is building exports.

Glass in Architecture

Advances in glass production have been dramatically demonstrated and exploited in modern architecture.

Since the late 19th century the size of windows and the proportion of glass to solid in building facades has increased greatly until the 'glass facade' has become commonplace.

In the last 20 years the 'all glass assembly' of components has further exploited the potential of glass in architecture.

Close liaison between representatives of the glass and building industries has helped bring this about. Viridian has produced an excellent book, 'Architectural Glass Specifiers Guide', which covers the many and varied types and uses of glass currently available.

Scarcely any building is built today that does not use numerous products of the glass industry where they can be seen in the finished form.

GLASS

A city building facade in which the glass and the frames holding it are used to form a striking pattern.

An all-glass assembly used to minimise the separation of outdoors and indoors in a department store.

OTHER TYPES

Diffuse reflection glass is a lightly textured glass designed for use in picture frames and instrument glasses to reduce annoying reflections.

Mirrors are made by a continuous silvering process on large sheets of glass. The glass is cleaned, tinned, silvered, then coppered, painted and baked to protect the silver coating.

Lead glass. For protection from dangerous rays such as X-rays and gamma rays, lead oxide is added to glass and a sheet approximately 25 mm in thickness is used.

Photochromic glasses. Development of these glasses, which darken with exposure to light and recover when the light is removed, have not yet reached the building industry but are in use for optical lenses. It seems likely they will be ideal for many building uses if costs are not prohibitive.

CARE AND CLEANING OF GLASS

Whilst glass is extremely hard and resistant to many forms of chemical attack, it does require careful treatment, especially on building sites.

The surface can be etched by cement wash, mortar or slurry if left and allowed to dry, even after completion of the building. Removal of such material must be by washing as soon as possible in order to avoid scratching.

Torn protective covers on pre-glazed units often conceal such droppings until stripped off. Replacement of the glass is the only way to overcome such damage.

Similar damage can be caused by some adhesives of alkaline nature which are used to attach suppliers' names on warning signs.

Reflective glass is now commonly used both for heat control and external effect.

GLASS

VARIABLE PROPERTIES OF GLASSES

As the sizes of glass sheets used in buildings increase, the need to understand the physical characteristics of selected glasses also increases.

Density and stresses. The density of normal window glass is about 2.5 (i.e. equivalent to aluminium) but lead crystal approximates 3.1 and some optical glasses equal 7.2.

In structural properties, sheet glass is very strong in compression. When it breaks it does so in tension. Toughened glass will withstand impact loads greatly in excess of sheet glass capability.

Laminated glasses can prevent dangerous shattering in accidents.

Thermal Properties

Thermal expansion is important as a factor in the resistance of glass to heat shock, and in determining the stresses set up in windows under alternating heating and cooling. In glass-to-metal seals, the expansions of glass and metal must match closely over the range of temperatures in which glass cannot yield to stress.

Thermal endurance or the ability to resist temperature contrasts is important in such items as cooking utensils and some industrial building situations.

Special glasses such as Pyrex have been developed that can resist shock of 300 deg C. Normal window glasses can tolerate approximately 130 deg C, which generally is adequate.

The thermal conductivity of glass varies considerably according to its constituents, some glasses being three times more conductive than others.

The range of values is from 0.0028 to 0.0078 calories per centimetre per degree centigrade per second. For most building purposes, manufacturers can provide more meaningful values for each type marketed.

Light absorption and transmission

Of the white light falling on glass at right angles to the surface, some is reflected, some absorbed, and some emerges at the other surface.

The degree of absorption varies with different wavelengths, so the emergent light is different from the incident light.

The extent of the absorptions depend on glass composition and thickness for building situations. Glasses very low in iron content transmit ultra-violet light; a high ferrous iron content cuts off both ultra-violet and heat radiation; nickel oxide content obstructs visible light but transmits ultra-violet.

Glass manufacturers should be consulted where large quantities or sizes of glass are contemplated, as their recommendations and advice can affect greatly the whole specification and detailing of glazed openings.

Obviously, consultation during the early stages of building documentation is most desirable.

An indication of the strength of toughened glass is given by its wide use in squash courts.

Surface treatments

The surface of some sheet glasses can be worked to produce permanent designs or patterns by etching, sand blasting, cutting and chemical deposition by hand and machine processes. Manufacturers should be consulted regarding such treatments.

SUMMARY OF GLASS

Structural strength. Sheet glass has very good compressive strength and reasonable tensile strength but is brittle in its normal sheet forms.

The ability of sheet glass to resist wind pressure (one of its major structural requirements) is very much a product of the in situ sheet sizes and the thickness of the glass sheet and wind velocity. Tables for various glasses are available from manufacturers.

Laminated and toughened glasses are usually of greater thickness than are sheet glasses and have much greater structural strength and resistance to cracking.

Manufacture and availability. Glass sheet sizes vary with regard to manufacturers, thicknesses and types. Sheets are now limited principally by furnace sizes and site handling difficulties. Thin sheets up to 5 mm thick are usually restricted to panes approximately 1800 mm x 1800 mm. Normally, sheet glass for windows is available from stocks held by distributors and may be cut to size using diamond or hard metal cutters. Sheets of 3 mm are common in domestic work.

Toughened or safety glass must be made to order and to size and cannot be worked or altered after leaving the factory. Large sheets approximately 4 m x 2 m are frequently used. Sufficient lead time must be allowed between ordering and delivery.

Water and freeze/thaw effects. Glass is particularly inert and unaffected by water in building situations. Being non-absorbent glass itself is not affected by freeze/thaw but can be cracked by expansion or contraction occurring in adjacent members in contact with it, especially if a point of pressure is exerted on a cut edge.

The major liquid affecting glass is hydrofluoric acid, which can etch the surface. Water-borne cement slurry can also mark the surface if left long after drying.

Temperature effects. Temperatures around freezing increase the brittleness of glass. In fires, glass usually cracks because of temperature contrast, but some special laminated or wired glasses can resist shattering.

The conductivity and heat transmission of glass is an important element affecting comfort conditions where large windows are used. Glass areas usually have the highest thermal conductivity factor of all walling materials. This factor can be improved by use of double glazing, heavy glass or the specially developed environmental control glasses.

Ultra-violet radiation. Glass itself is not affected by ultraviolet radiation, and glass colours are usually permanent, but glass influences the passage of ultra-violet and other light waves through it. Light waves striking the glass surface are refracted, some reflected, and some absorbed, depending on the colour and other constituents of the glass.

Special glasses have been developed to allow the passage of ultra-violet waves, and others can obstruct these effectively.

Electrolytic or other special effects. Glass is a good electrical insulator at normal temperatures, and is non-reactive with metals, minerals or other building materials.

The unique quality of glass is its ability to resist water penetration and wind effects, yet admit a very high percentage of light.

Acoustic qualities. The smooth, hard surfaces of glass tend to reflect sound. Because of the thinness of the sheet the unit mass is low so that glass areas are not very effective barriers to sound. Increasing the thickness of glass can improve acoustic insulation for closed windows, and double glazing with an air gap can be even more effective.

GLASS PRODUCTS

PHYSICAL PROPERTIES OF GLASS FOR BUILDING
By courtesy of Pilkington ACI Ltd.

STRUCTURAL

Density	2500 kg/m³
Specific gravity at 21°C approx.	2.5
Compressive strength for 25 mm cube	24.8 MPa
Tensile strength for sustained loading (See also percentage Fracture tabulation)	19.3 to 28.4 MPa

PERCENTAGE FRACTURE RISK WORKING STRESS TABULATION

Working stress for various flat glasses for a risk of not more than 1%	Sustained Load MPa	Momentary Load (3 seconds) MPa
Float glass up to 6 mm and all Sheet glass	15.2	40.4
Float glass over 6 mm	10.3	27.6
Rolled Patterned glass (depending on pattern)	6.9 to 9.0	20.7 to 23.4
Wired glasses	7.6	20.2

THERMAL

Linear thermal expansion coefficient	75 to 80 x 10⁻⁷ per °C
Thermal conductivity (K value)	1.05 W/m°C
Thermal transmittance (U value) for single glass, usually assumed as 6 for calculations	6.02 W/m²/°C for summer 6.42 W/m²/°C for winter

LIGHTING

Visible light transmissions for 6 mm glass with diffuse light	Sheet 85%: Rough Cast 80% Plate Float 85% Wired Cast 75% Translucent 70-85% dependent on pattern
Reflection loss—varies with angle of incidence of light	Approx. 8 to 10% for normal (90%) incidence

ACOUSTICAL

	Decibels	Db
Mean noise reductions for: Single sheet glazing	4 mm-25 8 mm-29 12 mm-31	6 mm-27 10 mm-30
	Db	Db
Laminated glass	6 mm-29	12/14 mm-34
Double glazing for sashes of glass/airspace/glass in mm	6/12/6-29 10/100/6-39	10/12/6-31 6/206/6-41

Standards

AS 1288-1994 — Glass in buildings
AS 1735.2 Lifts — Escalators & moving walks
AS 2208:1978 — Safety Glazing materials
BS 4031:1966 — X-ray protective leaded glasses
BS 5051 — Security glazing
BS 5544:1978 — Anti-Bandit Glazing
BS 5713:1979 — Double Glazing

Reference Books

McLellan & Shand; Glass Engineering Handbook.
Viridian New World Glass; Architectural Glass Specifiers Guide

GLASS PRODUCTS

Glass fibres

Because of the viscous nature of molten glass it has always been possible to draw a fine filament of the glass out of the melt. Experiments in spinning such filaments on to a wheel resulted in long, continuous glass fibres which, however, were too coarse to permit fabrics made of them to be folded, prior to the present century.

Numerous patents designed to industrialise the drawing of glass fibres were taken out prior to 1914, and during World War I the Germans produced coarse glass fibres to replace imported asbestos and other high-temperature insulation materials. By 1930 effective processes were in operation in both Europe and the USA, and markets for the product were being developed.

During the 1930s, two large American glass manufacturers entered the field and combined forces in 1938 to form the OwensCorning Fiberglas Corporation. Together they developed new processes and products.

After World War II, a vast new field of application was found for glass fibres as plastics reinforcement, glass decorative fabrics, medical applications and insulation.

Glass fibres are inorganic, incombustible, non-absorbent, non-hygroscopic and chemically stable. The diameter of most of the fibres used is so small in relation to their individual mass and length they can be bent to a radius several times the diameter of the fibre without breaking. The tensile strength greatly exceeds that of sheet glass and also of any other commercial fibre.

Fibres are available in numerous forms which are further processed to produce materials suited to many purposes. The major building uses are as follows:

Glass wool is a resilient fleece-like mass processed to produce batts. blankets, boards, blocks, pipe insulation and cements for many thermal and sound insulation purposes. Care is needed in handling glass wool products as they can be a skin irritant, but no evidence exists they are a health hazard.

Glass textiles. Continuous filament fibres and staple fibres are produced and formed into yarns, mats and woven fabrics for a great variety of uses including pipe insulation, the base for waterproof roofing felts, reinforcement for plastic roof sheetings, etc.

Foam glass. By packing glass granules into moulds with a material that will produce gas bubbles at the appropriate temperature, a lightweight building block or board can be produced which can be sawn and shaped. Used as rigid insulation, this material is rot and vermin proof, with a very low water absorption. This foam glass has been used as the insulation core in several sandwich-type claddings for buildings.

Glass-reinforced cement. The use of glass fibres as reinforcement for the tried and versatile mix of silica sand, water and cement is a comparatively recent (1970s) innovation which shows tremendous potential for building products and applications. In the wet state it appears like a hairy cement mix. The glass fibre content used is about 5% by weight of the total. The glass fibres have to be of a specially alkali-resistant type called Cem-FIL which have been developed by Pilkington and subjected to extensive testing over a period of 10 years, with very good results.

The advantages of GRC are numerous, some of the most relevant being:

* The non-corrosive quality of the reinforcing fibres.
* The ability to produce elements much thinner and lighter of weight than is possible with steel reinforcement (10 mm is a typical application).
* Ease of forming, shaping and cutting.
* Weather resistance.
* Non-combustibility.

The material can be premixed and shaped in a manner similar to concrete casting, pressing and extruding. More commonly, the mixture is made by a spraying process using measured amounts of raw materials. In some cases a further step to improve finished quality is called spray dewatering.

The whole production process is controlled by patents and a system of licences issued through Pilkington's. This enables development to avoid the effects of poor quality control and resulting defects or failures which usually beset the introduction of new products.

Australian production and distribution of glass fibres is in the hands of ACI who have technical staff and literature available.

Frameless glass doors are made possible by the use of toughened glass.

Glass fibre batts being installed as insulation in timber framed walls.

A variety of glass fibres are produced in chopped, strand and woven forms for various applications.

Cem FIL chopped fibres, cement and sand being applied by a gun process to produce GRC over a mould. Note the hairy nature of the surface produced on the gun side.

Porcelain enamel. This material is really a form of glass applied to another base material at high temperature so that the base and surface bond together.

Cast iron porcelain enamel (CIPE) sanitaryware is a longestablished example of this in building. Other uses such as enamelled facing sheets have periodically been popular.

The facing has the hard weather-resistant properties of glass and can be produced in various colours.

Electrical goods. Glass is a very important component of electrical goods used in building, from the housing around lightproducing electrical elements to the insulation around wiring, from lead crystal chandeliers to cheap pressed glass lampshades.

The range and applications of glass uses in this field alone is indicative of the tremendous importance of glass in the building industry and the wealth of imagination that has been applied to its exploitation in recent years.

Glass bricks, developed in the 1930's, are now available in a range of shapes and sizes, and have been accepted as fire rated in some situations. Details regarding installation etc. are available from major manufacturers and suppliers. They need to be installed by tradesmen skilled in the special techniques required.

Decorative Glasses

Layering, fusing, etching, spray finishing and other techniques using coloured and patterned glasses is opening up a new field of decorative design by imaginative crafts people, while others carry on the ancient lead joined stained glass window tradition.

Standards

AS 1288	*Glass in Buildings*
or NZS 4223	*Glazing in Building*
AS/NZS 4666	*Insulating Glass Units*
AS 1530.4	*Fire tests on materials, components & structures*
ICANZ-2000	*Glasswool & Rockwool. Industry code of Practice*

Acknowledgements

John Young - DESIGNING WITH GRC
Architectural Press
Pilkington ACI Ltd, ACI Fibreglass;
Dow Corning (Aust) Pty Ltd, for illustrations.

Extensive use of glass bricks demonstrated at an ACI factory built in 1940.

Individual glass bricks and a screen wall suggest further useful applications.

Some obscure glass patterns made by the drawn and pressed process.

Laminated bandit glass resists deliberate attack.

Rough Cast   Kosciusko

Spotswood   Bullion

Cosmos   Mauresque

CHEMICALLY BASED PRODUCTS
History; Asphalt; Bitumen; Resins; Mastics and Caulkings; Paints;
Tarnishes; Plastics; Thermoplastics; Thermosets; Plastic products; Summary.

INTRODUCTION

Compared with other industries engaged in the production of building materials the chemical industry, at under 200 years of age, is very young indeed. However, in that time it has succeeded in invading directly or indirectly every aspect of building. Many of its products compete with traditional building products, and its aggressive marketing practices have forced the reassessment and improvement of older materials and methods.

Without chemical products all 20th-century industry would quickly grind to a halt. Much of our technology depends on it for raw materials, lubricants, solvents, and dozens of other important items.

Some building products that are, or were, available naturally have been duplicated chemically by the industry that now dominates in their production. This applies particularly to asphaltic, caulking, sealing and painting products. Consequently, these materials have been grouped under this general section as chemically based products.

Aims and history

The chemicals industry takes a comparatively few fundamental raw materials (e.g. coal, salt, limestone, sulphur, water and petroleum) then combines them and recombines them in ways that are new and useful. This is called 'synthesis'.

The aim of the chemicals industry is to produce new products - and better products - economically. This is the basis on which it has developed since the beginning of the 19th century.

Nicholas Leblanc showed how to produce soda ash as a cheap alkali in France in 1790. This discovery has been called the most important industrial chemical process of the 18th century.

Leblanc unfortunately suffered confiscation of his patents and factories during the French Revolution and never recovered. After his death in 1806, however, his products were needed by soap and textile manufacturers. In England, chemical manufacturing industries sprang up and soon had large quantities of byproducts which were an embarrassment to them. As a result uses were developed for the by-products, often by further chemical treatment.

Along with these developments which were based on minerals came the introduction of chemicals obtained from living organisms, animals, plants, etc., all of which contain carbon. Thus was initiated the academic subdivision of inorganic chemistry based on minerals, and organic chemistry based on carbon compounds.

In 1856 the production of commercial substitutes for organic chemicals commenced, pioneered by the Englishman, Perkin. Most of these materials were derived from the benzene in coal tar.

European chemists became prominent innovators, and some of the world-renowned chemical manufacturers such as Bayer were well established by the end of the century.

Up to World War 1, Germany led the field in utilising the byproducts but was overtaken as a producer of organic chemicals by the United States of America in the years between the wars.

The distillation of petroleum products commenced about 1856 to produce lamp oil, and it was another 40 years before gasoline was used for automobiles.

This petroleum and gas distillation now provides the basis for six major groups of end-products. Bitumen, synthetic rubber, plastics, textiles, detergents and agricultural chemicals, as well as the familiar fuels and lubricants.

Whilst some products are now well proven in use, many newer ones lack the performance records needed for products to be used as major components in building.

At present they are generally confined to minor components or surface finishes, but no doubt some products will develop into competitors for the more traditional materials.

ASPHALT AND BITUMEN

Asphalt is the semi-solid, sticky, black-brown residue formed by the partial evaporation or distillation of certain petroleums. It is found in its natural state in some locations and has been used for construction purposes for thousands of years.

Natural asphalts available for building purposes are hard and rock-like and need to be heated to liquefy and remove water, gas and other volatile materials before use. This is done usually in a large wood- or gas-fired 'kettle' on the building site.

Methods of refining asphalt from petroleum were developed in the 20th century and asphalts now come mainly from such sources. Petroleum asphalts are usually called 'bitumens' and are refined to the desired consistency for working requirements. The terms 'asphaltic' or 'bituminous', as used in building, are now virtually interchangeable.

Bitumen was the name used by the Romans for the natural hydrocarbons such as natural asphalt. Its use now in buildings usually relates to the tar-like waterproofing mixture of certain hydrocarbons used in the production of asphalt, roofing felts and dampcourses.

The road-building and construction industries use large quantities of these materials. They mix readily with gravels and sands to produce flexible paving surfaces that are also waterproof and make ideal roads and pavements. The self-adhesive nature of the materials enables jointless surfaces to be laid in many situations.

Mastic asphalt is the term commonly used to describe a mixture of bitumen and crushed rock aggregate that can be applied in a hot fluid condition for paving or waterproofing purposes.

Hot asphalt or bitumen is liquid and often used as an adhesive coating for other materials on flat roofs. When cool it gives an elastic waterproof base. It is also applied hot to vertical wall surfaces as waterproofing.

Bituminous Roofs

Many products are made using fibres saturated in bitumen for roofing purposes and these are often fixed in place using bitumen adhesives either hot or cold. Sands and gravels are often incorporated into built-up roofing membranes.

One of the big advantages of bituminous products is that they can readily be joined or patched if cracks occur or holes have to be cut in them; however, it is difficult to determine on buildings if the patch is as secure as the original job.

The major disadvantage in roofing situations is that ultra-violet rays affect asphalt and bituminous products, making them hard and brittle after years of exposure, and the eventual breakdown of the roof is inevitable.

The life-expectancy of such installations depends very much on the ultra-violet light intensity of the geographic location, the thickness of the built-up roofing membrane and the nature of the fibres or gravels built into the sheet material.

As Australian sunlight has a very high ultra-violet content, these roofs are far less reliable in this country than in many other parts of the world.

CHEMICALLY BASED PRODUCTS

Bituminous products are also sensitive to temperatures generated by direct sunlight. Because of their black-brown colour they absorb much of the sun's heat and become soft on very hot days, causing creeping and thinning in exposed locations. Any loading or foot traffic in such hot conditions may squeeze out, penetrate or remove some bitumen, leading to weak spots appearing in the waterproof membrane.

These defects can only be partly overcome by protecting the bitumen from sunlight with a layer of reflective gravels or tiles.

Bituminous felts are impregnated fabrics, either felted or woven fibres of various kinds such as jute (hessian), or glass fibre, and are mainly used in producing waterproof roofing membranes.

The glass fibre materials are comparatively new. As they are not affected by water and have extremely high tensile strength, their life potential is superior to that of the older asbestos- and jute-based materials.

Bituminous paints are used to protect metals from corrosion where their very strong, dark colour is not disfiguring. Protective coatings are also factory applied to some products such as aluminium dampcourses.

Some solvents are used to produce cold liquid bitumens which harden as the solvents dry out. This is the basis of most bituminous paints, cold adhesives and patching compounds.

RESINS

Resins are very important basic ingredients in many products. The various types of resins used are often reflected in the names of resulting products.

It is now generally recognised that resins, plastics, rubbers and fibres are part of one large group of chemical components and it is possible to change any one member of the group into another. In the process, combinations sometimes occur between natural and synthetic products to achieve special qualities.

Natural resins are obtained as gum exudations from certain pine and fir trees when damaged. These resins become harder and more insoluble with age, and in some localities are found as deposits in the ground where forests once stood.

The kauri resins from New Zealand are still highly valued and in many instances are known to be over a thousand years old. Older civilised areas have used resinous materials for paints and lacquers from early historic times. The hardest of the natural resins is amber, which is used as a semi-precious stone. Natural tree gums are usually soluble in water and are often used in glue manufacture; but the resins are changed chemically, are not water soluble but combine with oils and organic liquids. Natural varnish resins are transparent, translucent, brittle and of brown or yellow colour. By heating, most insoluble resins become readily dispersible in oils.

Dammar, a hard resin usually of pale yellow beads which melt at about 140°C, comes from Malaysia and Indonesia. These are the base for many rapid-drying lacquers.

Copal resins are an important group for varnish making and include the New Zealand kauri resin as well as others from Africa and South America. They melt at about 150°C.

Synthetic resins possess most of the physical properties of natural resins and in addition have many unique properties of their own. They are widely used in the industrial coating (paint) industry, and also for adhesives, binders and textile impregnants.

Following are some of the resins commonly used in building products. They are given here in alphabetical order.

Acrylic resins are derived from acrylic and methacrylic acids which generally involve cyanide and alcohol in their syntheses. Various combinations produce materials ranging from almost liquid to very hard. The clear sheet form is used widely in aircraft windows, and other forms are used in paint manufacture.

Alkyd resins are reaction products of alcohols and acids, the name having derived from the two components by changing 'cid' to 'kyd'. These date from the mid-19th century.

Paint products based on these resins are usually of the enamel type, having excellent metal adhesion and heat resistance. They form the basis of many coatings for industrial products such as cars, refrigerators, etc.

Epoxy resins are phenolic resins which generally can be cured to hardness quickly by chemical reaction. This is the basis of the effective two-pot-type coatings and glues which produce very hard, flexible and durable surfaces. Similar epoxies which are supplied in one package form and rely on air curing do not produce the same degrees of durability.

Melamine formaldehyde resins are similar to urea resins but produce better heat and chemical resistance, with outdoor durability and colour retention. They are used widely for automotive and industrial finishes.

Phenolic resins. About 1928, phenol formaldehyde resins were produced which found wide acceptance in the varnish field. Durable finishes have been made from these resins. They are sometimes mixed with alkyd resin products.

Polyester resins are closely related to alkyd resins and are used in various forms such as with glass fibre to become surface coatings or textile fibres.

Polystyrene resins are sometimes called vinyl benzene, which relates them to the vinyl resins. Expanded polystyrene products now provide very efficient insulation materials for building uses.

Silicones are based on a silicon-oxygen chemical skeleton instead of the carbon chains found in most resins. These develop a resistance to heat not found in carbon-based resins. Various polymeric structures of silicone can be made in order to produce resinous, liquid or rubber-like materials. High temperature resisting paints and enamels can be made from these bases, as well as mastics and sealants.

Silicones are chemically inert and are affected only by strong alkalis and hydrofluoric acid.

Thermoplastic and thermosetting resins. Physically, it is convenient to classify synthetic resins by their solubility. There are resins that remain soluble (thermoplastic) and those that are initially soluble but become insoluble and infusible under the action of heat (thermosetting).

Urea formaldehyde resins are chemically neutral, of very light colour, can be mixed with any pigment or blended with other resins to maximise the virtues of each. These products develop hardness and gloss when heat is applied.

Vinyl resins are generally of a rubbery nature and can be developed into adhesives used in heat bonding and paints (PVA, or polyvinylacetate).

Another form (PVC, or polyvinyl chloride) is fire resistant and suited to insulation for electrical wiring, etc.

MASTICS AND CAULKING MATERIALS

Many materials are now used in building to seal junctions between components. The use of putty as a weatherproofing seal for glass in timber sashes is a long-established example. This is composed of whiting and linseed oil.

On a larger scale, asphaltic materials have been used for joints between concrete slabs in roadwork to allow for shrinkage and expansion. The variety and uses of such materials have increased greatly with the introduction of more steel, aluminium and precast concrete in buildings and the desire to create buildings of an assembly of 'dry' materials.

The need to weatherproof expansion and construction joints has developed a sub-industry serving the purpose.

These materials are generally grouped into the category of mastics and caulkings. Their primary requirements are to have suitable degrees of flexibility to adjust to building movements, and to retain this quality for lengthy periods without squeezing out or otherwise becoming misplaced through tension or gravity.

Some of these materials are viscous and adhesive; others are flexible and dry. Obviously, the base materials can vary considerably. They include:
asphalts and bitumens
rubbers and neoprene
polysulfides
sponge-type materials
brush-type materials
butyls
silicones
putties
These materials are now products of the chemicals industries and some are formulated for very specific functions.

Some demonstrate an extended application of uses proved already in the automobile industry.

CHEMICALLY BASED PRODUCTS

PAINTS AND VARNISHES

Many materials have been used to apply surface coatings to buildings, usually by brush application. The purpose has been to prolong the life of components, to improve waterproofing performance, and to decorate.

Locally available products of mineral, animal and agricultural origin became recognised as effective for these purposes and in many cases contributed largely to the architectural character of old-established settlements.

Lime wash for mud and plastered walls was one of the old and widely used waterproofing paints, whilst clear and pigmented oilbased varnishes have been known for centuries. The development of weather-resistant paints for timber was probably a much later innovation.

Paints generally consist of basic pigments in a 'vehicle' or 'medium' to create a spreadable product. The vehicle can be a varnish, oil, water or other liquid that will evaporate on application.

Linseed oil has been very widely used for a long time as a major vehicle, with natural turpentine and, later, mineral turpentine and white spirit as a thinner and drier.

Stainers are used with some clear finishes to obtain required colours, whilst pigments are opaque and generally more weatherresistant.

Paint manufacture is now part of the vast chemicals industry and is closely allied to the plastics industry in this regard. New materials and methods are continuously researched and tested and the results communicated across the world; consequently, the relative merits of many products are not proved by the long-term job performance criteria of most building materials, but more by short-term exposure and laboratory testing. These methods are appropriate for applied surface finishes such as paints, varnishes and stains which need to be renewed or recoated regularly.

Pigments are finely ground insoluble white or coloured powders obtained from natural earth colours or chemical processes.

White pigments that are widely used are white lead, zinc oxide and titanium oxide.

Coloured pigments are numerous; they include chrome yellow, yellow ochre, ultramarine, red lead, iron oxide, umber, burnt sienna, carbon black.

Metallic powders of copper, zinc, lead and aluminium are also used for special effects and protective coatings.

Combinations of pigments can produce an almost infinite number of colours.

Varnish is a homogeneous organic composition which, when spread thinly in liquid form, dries on exposure to air or heat to a hard transparent film which decorates and protects the surface to which is is applied. It may incorporate dyes or be used as the vehicle or binder for opaque pigments, as in some paints.

Varnishes usually contain resins or gums (which provide the hardness and lustre), oils, solvents and driers.

Exterior varnishes are rich in oils, whilst harder interior varnishes have more resin in them.

Water-thinned varnishes developed rapidly during World War II when non-flammable binders were needed for paints that were to be shipped in concentrated form and then thinned with water. This need led to many new types of paint being produced in the post-war years.

Stains are sometimes used directly onto woodwork to colour it without obscuring the natural grain, as a preliminary to varnishing.

Water stains are solutions of dyes of various colours in water and need to be varnished or waxed in order to waterproof them. Spirit stains are dyes that are soluble in methylated spirit and sometimes contain small quantities of semi-transparent pigments such as burnt sienna or vandyke brown.

Oil stains are very thin consistency oil paints, containing only a small proportion of pigment.

Creosote is often used for exterior staining of woodwork, where it also acts as a preservative. Other finishes cannot be applied readily over creosote as the creosote will usually bleed through paints unless first coated with an aluminium paint.

Priming paints. Paints used for the first or priming coats have a very important bearing on the life expectancy of the subsequent coats. They provide the initial protection to the surface from the penetration of moisture and form a base for the next coat. For timber, it is essential they should penetrate into the surfaces as much as possible, therefore they are a little thinner than subsequent coats.

For exterior woodwork, primers are usually oil-rich and pigmented with white lead blended with a proportion of red lead, to produce a pink colour. Water paints should not be used on woodwork, with the exception of modern acrylic paints.

For interior work that is not exposed to moisture, primers are made with lead-free pigments such as zinc oxide, lithopone or titanium white.

Primers for iron and steelwork should be made with a large proportion of rust-inhibiting pigments such as red lead, iron oxide, chromate of zinc or zinc dust. Galvanised steel needs special treatment, as outlined later.

The special needs for painting galvanised steel indicate how important it is for the primers to be compatible, not only with the base material but with the finishing coat as well.

Water paints. This term applies traditionally to a range of paints made from finely ground limestone pigments in powder form, some with cement added and mixed with water on site to produce economical surface finishes.

With some of the newer plastic and acrylic paints, brushes are washable in water soon after use, but once the film hardens (which it does quickly) water does not affect the paint. These paints, however, are not to be considered as water paints. Acrylics are now being used externally with very good results.

Modern synthetic paints have almost completely superseded the old-style water paints.

Distemper, or Kalsomine, was probably the best-known of these finishes. Their life expectancy was reasonable, but they were non-washable and easily marked. Other coatings cannot be applied successfully over these coats.

Some modifications to the simple kalsomine were made to improve adhesion by adding powdered casein as a binder. Washable kalsomines were called 'oil-bound water paints', which were usually supplied in a paste form then thinned with water for interior use, or with petrifying liquid for exterior use.

Cement paints were white cement and pigments in powder form which had to be used within an hour of adding water on the site. They were effective on exterior plaster or masonry.

Bituminous water paints have been produced which have waterproofing properties after drying, but they are limited in colour to the red-brown-black range and should be used only on masonry or similar absorbent surfaces, not on metals.

Mediums or vehicles used for paint manufacture are the liquids that act as binders for the pigments.

Raw and processed linseed oils have a long history of use, especially for coatings on to timbers. Other vegetable oils such as safflower are now used as well, but linseed retains its reputation as the best available for priming paints applied to timber.

Turpentine, both natural and mineral, and white spirit are used with the oils to obtain suitable consistency. Dryers such as terebine are used in small proportions to promote drying action.

Varnishes incorporating these with soluble resins are also used as the vehicle for some paints, especially the hard, glossy enamels.

Ready-mixed paint

Up to World War II it was not uncommon for painters to mix paints on the job, in order to meet specific requirements, using ground-in oil pigments or powders and adding their own colours and vehicles. This procedure required a high level of knowledge and skill so that one could be sure of good results.

Now, paint manufacturers have a wide variety of prepared paints and colours to suit normal situations. It is possible to select colours from a chart and have colouring pigments added by the distributor to match the chart. Basic colouring pigments to tint or alter ready-made colours are also available in tubes so that almost any colour is achievable.

Where consistent colour and quality are required, the factoryproduced ready-mixed paint straight from the can is the safest to apply.

Gloss painted timber members showing two degrees of paint deterioration due to a 10-degree variation in the angle of exposure to the sun. The nearer timbers receive marginally more sunlight than the farther ones and show considerably more deterioration.

It is still necessary, however, to choose an appropriate type of paint for the situation, as the material over which paint is applied can react unfavourably with the ingredients. Hence the numerous types of priming paints that prepare surfaces for coatings to follow.

Enamel paints are so called because of the high gloss finish suggestive of the enamel of vitreous glazes. They are made by finely grinding the selected pigments in a varnish medium. The nature and properties of the finished paintwork depend on the type of varnish used.

Traditionally, the exterior enamels were slow-drying, needing 12 hours or more, whilst interior enamels were available which dried much more quickly.

Recent innovations in synthetic base enamels have reduced considerably the drying times. Enamel paints have good abrasion resistance and toughness. Used internally they retain their gloss well; used externally, they need recoating after three to five years, gradually losing their gloss during that time.

Flat paints (not to be confused with water paints that produce a similarly flat finish such as kalsomine and oil-bound water paints). These paints are really flat-drying enamels, except that the pigments are less finely ground and the vehicle contains less varnish and more volatile solvents. They are widely used internally but are not suitable for external use.

Anti-corrosion paints are produced to prevent iron and steel from rusting and are effective if maintained regularly, as on major steel bridges and other engineered structures.

Red lead in linseed oil was traditionally used for this purpose, but lead based paints are now banned as public health risks. Zinc chromate and heavy bodied zinc paints are available as replacements, but usually need top coats to conceal their harsh colours. Epoxy based paints are now also widely used. In situations involving acid exposure chlorinated rubbers perform better than the epoxys. Bituminous paints are useful in some situations, such as eaves gutter interiors, but restricted due to their limited colour range and tendency to bleed through other coatings. The most recently developed polysiloxane protective coatings (PSX) are claimed to outperform three coat epoxy/polyurethane systems.

Metallic paints. These paints are prepared by mixing minute flakes of aluminium, copper, bronze or gold in a quick-drying oil varnish.

Aluminium paint provides a brilliant reflective coating for light and heat and is especially suited to use on metals. It is also very efficient as an undercoat before applying light-coloured finishes over dark paint, and as a sealing coat over creosote to stop it from bleeding through subsequent coats.

The bronzes are suited to painting metals, and the golds are used for highlighting internal decorative detail.

Paints for galvanised steel. Galvanised steel is often painted, and frequently the paint film peels off within 12 to 18 months. This problem has led to considerable research and product development by paint manufacturers. It has been found that the basic cause is poor adhesion of paint to the smooth galvanised surface.

Zinc used in galvanising is a reactive metal and reacts with many paint vehicles, causing the paint to lose adhesion or embrittle. For good performance on galvanised steel, the first requirement is that the paint vehicle be non-reactive to zinc. The second important factor is surface preparation.

For high-performance applications, light abrasive blasting of the zinc surface is the best technique for guaranteed coating performance. Surface priming systems are more commonly used in building situations, some of which are as follows:

Vinyl etch primers need spray applications and careful control.

Two-pack epoxy/polyamide primers are suitable for use with almost all paint finishes.

Acrylic latex primers are water-based, easy to use, non-toxic, and provide good adhesion. These give best results if coated with exterior acrylic finishes.

Zinc dust-zinc oxide pigmented primers in vegetable oil binders give good results under conventional alkyd enamels, which are otherwise reactive with zinc.

Fireproof paints. These paints are made for use on woodwork and similar flammable materials. They contain ingredients such as asbestos, borax or other fire retardants but are not usually acceptable to fire authorities as giving an acceptable fire rating.

Plastic paints. From 1940 onwards, as the paint and plastics industries became closely allied, many new types of paints were available, some of which were called plastic paints for publicity and marketing purposes.

The paint industry was in many ways the forerunner of the plastics industry, in that their uses of organic resins, oils, solvents, etc. are common and the products are the result of complex chemical processes which have been understood only in the last hundred years.

The term 'plastic' is too general to be useful in the context of paints and the term 'latex' is preferred.

Latex paints are water soluble, sticky, and are made from a solid resin. They can be of flat, low gloss or semi-gloss finishes. The vinyl resins were the basis for early latex paints, especially PVA (polyvinylacetate). These are now superseded.

Polyvinylchloride (PVC) blended with PVA can produce excellent quality coatings for steel where in contact with water, as in dam sluice gates, etc.

Breathing paints. Whilst paint is usually intended to keep moisture out of the item coated, in some situations it is desirable for the moisture within to get out through the surface. This is often the case in newly constructed concrete or masonry walls that contain water during the wet construction process.

In these situations paints that will allow evaporation to occur through them are necessary. They are usually the old water-based varieties or some of the flat paints.

Sealers and gloss paints should not be used on both sides of such walls until they are thoroughly dry, otherwise blistering can occur from the hydrostatic forces in the wall.

Sealers are applied for various reasons, usually -

(a)   to enable finishing coats to dry out effectively and avoid being affected by reactive elements in the base material;

(b)   to reduce the suction of the base and ensure an even finish for the final coats.

Sealers, therefore, are a special type of primer. They are usually pale or clear coloured and are prepared from latex, casein or silicone bases.

Waterproofers are similar to sealers but are intended to keep water out of masonry and similar porous materials. Bituminous paints, latex and silicones are used for this purpose.

CHEMICALLY BASED PRODUCTS

Acrylics

Acrylic paints have become a major component of the range of so called house paints. They are easy for amateurs to use, have good opacity, and brushes are cleanable in water.

Textured Coatings.

Paint on textured coatings have been developed for both interior and exterior uses. Surfaces around swimming pools, where slipperiness must be avoided are a prime test for these paints which are quite new. Slip resistance is achieved by having hard granular materials embodied in the vinyl based vehicle. Time will show how well they perform.

External coatings for concrete and masonry can be built up in two or more layers to provide a lightly textured and coloured finish, such as 'Granosite'. These paints are usually applied by specialist tradesmen who are skilled in the on site preparation and application. As with all painting and surface coatings, preparation of the surface is as important as selecting the most suitable coating.

Durability

Obviously, surface finishes exposed to weather deteriorate more readily than internal finishes, and a life of from three to five years is usually considered reasonable for external paint on timbers or metals.

Internally, the life of paint depends largely on the atmospheric conditions encountered. Steam, smoke, moisture, regularity and type of cleaning, all affect paint life, but cases of paint lasting 20 or more years internally are not uncommon.

Externally, the degree of exposure to sun, atmospheric pollution and rain will affect paint life; strong sunlight generally being the most damaging factor.

Dark paints in the red-brown-black range absorb more heat than do lighter surfaces and tend to deteriorate, crack and peel more quickly. Reds seldom hold their colour well in strong sunlight.

Durability is influenced very much by the vehicle used, hence only materials recommended by the manufacturer should be added to paint.

Pigmented paint films wear much better than clear oil films alone, as the pigments reinforce the resultant film, making it dry harder and become more impervious to air and moisture.

Paints with glossy surface finishes tend to be more effective when exposed to weather than flat or semi-gloss finish paints, which are usually produced by decreasing the oil content and increasing the volatile component.

Climatic conditions vary due to geographic location and greatly affect surface finish performance. Popular local products are usually more suited to the conditions than exotic imports, and manufacturers' instructions should always be respected regarding preparation and application.

Paint application is usually by brush or roller to achieve the desired film thickness.

Spray painting

Most industrialised paint processes are done by spraying, but this method is not commonly used for buildings because of:

(a)   the variety of materials to be coated;
(b)   the unpredictable site and weather conditions;
(c)   the difficulty of achieving uniform coating thickness in such conditions;
(d)   the extent of the masking necessary.

Standards

AS A99-1959	Bituminous felt roofing
AS CA55-1970	Code for bituminous fabric roofing
AS 1160-1988	Bitumen emulsions
AS 1507-1980	Road tars for pavements
AS 2008-1980	Residual Bitumen for Pavements
AS/NZS-2310	Glossary of paint &painting terms
AS 2311-1992	The painting of buildings
AS 2312-1994	Protection of iron & steel
AS 3730	Guide to architectural paints

CSIRO NSB's

73	Epoxy resins
147	Paints
148	Paint systems - a guide to good practice

PLASTICS

Introduction

The term 'plastics' is now commonly used collectively for many man-made substances that do not exist in nature. In the 20th century the term 'synthetics' was applied to many of these materials and is still used, particularly in regard to textiles.

Most plastics are chemically produced organic materials based on carbon, but there is a growing number of exceptions on a formula basis, especially the silicones.

Plastics are high polymers; that is, made up of giant molecules composed of small repetitive units, assembled into large groups. As the name implies, they are plastic at some stage of production, and at that stage can be moulded by a variety of techniques.

Historical background

Many materials known today as plastics were discovered by chemists during the early 19th century - styrene in 1831, melamine in 1834, vinyl chloride in 1835, polyester in 1847, but their inventors did not appreciate the potential of their discoveries, which others later exploited. Twentieth century chemists adapted these very useful materials to many purposes.

The development of vulcanisation for rubber by Goodyear and Hancock, in America, was followed by the cold welding of rubber and the reclamation of waste rubber in about 1846 by Alexander Parkes, in England. This led to coating fabrics with rubber and the development of factory processes now used for plastics.

At the Great Exhibition of 1862, Parkes demonstrated his new material, called Parkesine, and by 1866 had established a company for its exploitation; however, faulty products led to the closure of the business.

Parkes's products were based on cellulose nitrate, which was also the formula used by Hyatt, in America, to produce billiard balls which could substitute for the traditional ivory balls. By 1870, a viable production of the materials called 'celluloid' had been achieved. Unfortunately this was a highly flammable product but was widely used for toys, sometimes with disastrous results.

It was not until 1907 that Leo Baekeland, in the USA, took out patents regarding articles made from phenolformaldehyde (which had been known since 1872) to establish the next commercially successful plastic, called 'Bakelite'. The colour range was limited from black to brown, and it had to be laminated with paper or cloth to overcome brittleness; however, it was moulded into many forms, from gramophone records to machine gears and radio cabinets.

Acrylic, developed by Chalmers in Canada (1929), was marketed by 1934 and widely used as a synthetic glazing material for the moulded shapes needed for aircraft during World War II.

Many other plastics developments were greatly advanced by Germany's activities in the textile industry and her desire to be independent of overseas raw materials. Britain's loss of the supplies of rubber from Malaya during World War II also forced action to develop substitutes.

In the post-war period, much effort switched to producing materials for specified properties. Applications within the conservative building industry were slow. The first noticeable intrusion of plastics was with the laminates used for bench and counter tops.

These and other products have proved their worth, the old stigma of 'synthetic' which had been used to resist the progress of plastics has been forgotten and new innovative materials are readily tried and tested. The building industry is now a major plastics consumer.

Parallel with the materials development came machinery and production processes to make plastics commercially viable. Some dates relevant to the building industry are listed here:

1820 - Hancock's rubber mill.
1845 - Bewley designed extruder for gutta percha tubes.
1872 - Hyatt Bros. patented injection moulding machines.
1879 - Grey patented first screw extruder.
1899 - Continuous film first made by casting on to a polished drum.
1909 - Baekeland granted 'heat and pressure' patent phenolic resins.
1915 - Synthetic rubber first produced.
1934 - First commercial production of Perspex.
1935 - Extruder for thermoplastics produced.
1937 - Polyurethanes first produced.

1938- Full-scale production of nylon.

1939 - First patent on epoxides. Commercial production
        of polyethylene.

1940 - PVC produced in UK.

1942 - Silicones produced industrially.

1950 - Teflon production on large scale.

1959 - Polycarbonates came on to the market.

In the late 20th century tension membranes were developed as sunshades and roof systems; made possible by the use of plastics, the skills of experienced sailmakers and computer aided design, to bring the forms of ancient tent systems into modern building design.

Advantages

The raw materials for plastics come largely from petroleum byproducts, coal and natural gas.

Water has little effect on plastics.

The density of plastics is generally low compared with that of metals. The heaviest plastics approximate to aluminium in density. Plastics can be developed to fulfil specific functions.

Very lightweight materials can be produced from foamed or cellular plastics, and these have very high thermal insulation values.

Strength-to-weight ratio is usually high, and in some cases exceeds that of metals.

Conductivity of plastics is generally low for both electrical and thermal effects.

Corrosion is variable, but many plastics are resistant to weak acids and alkalis. Strong acids or solvents may affect plastics and can lead to breakdown.

By reinforcing with other fibrous materials products can be achieved having the required qualities for a variety of applications.

Classification of plastics

Two major classes of plastics are generally recognised:

(a) Thermoplastics, which can be softened repeatedly with heat and
       will re-harden on cooling.

(b) Thermosets, which cannot be softened with heat once they
       are cured.

THERMOPLASTICS

Thermoplastics are composed of organic molecules consisting of very long chains of atoms (usually carbon atoms) in which the atoms are interconnected by strong (covalent) bonding forces. These chains are not interlinked and therefore will slip and flow on heating. Polyethylene, nylon and polystyrene are examples of this group.

At high temperatures thermoplastics are viscous liquids, and at very low temperatures they are brittle glasses. At intermediate temperatures they can be flexible, leathery solids or rubbers.

Most thermoplasts can be scratched by anything harder than an HB pencil. The hardest (acrylic) is comparable in hardness to aluminium.

Additives are widely used to improve certain properties and reduce costs of finished products, e.g.:

- Plasticisers increase flexibility in PVC.

- Fillers such as carbon black can increase strength and improve weather resistance.

- Flame retarders, stabilisers, pigments, etc. are also used.

Fabrication is usually by melting, then either by injection into a cavity mould or by extrusion.

Some thermoplastic materials

Acrylics (polymethylmethacrylate PMMA)

These materials are very stable, resist discolouration and degration from U-V radiation, can be coloured, translucent or opaque, and readily accept other coatings. Probably their most valuable quality for many applications is their transparency, with optical clarity, which has led to their use for lenses, light diffusers and other glass substitutes where moulded shapes or impact resistance are needed.

Acrylics can be cast, extruded, injection moulded and thermoformed, and are used in a great variety of building situations.

A large skylight made up from numerous acrylic domes and incorporating plastic sealing compounds.

ABS (acrylonitrite - butadiene - styrene terpolymer)

The properties of ABS depend on the relative proportions of the three constituents. In general, these materials have high tensile and impact strength, rigidity, resistance to heat, abrasion and creep.

Concentrated acids, alkalis, chlorinated hydrocarbons will attack ABS. The material can be moulded, extruded and thermoformed, offering a high gloss finish, and it can be metal plated.

Applications include the manufacture of telephone handsets, pipes and pipe fittings, door and window tracks and fittings, weather seals, etc.

Cellulosics

There are three basic kinds of cellulosic materials: cellulose acetate, cellulose acetate butyrate (CAB) and cellulose acetate propionate (CAP), all of which can be extruded, moulded and thermoformed.

Cellulose acetate is hard, stiff and tough, but it has poor dimensional stability with ageing and is attacked by alcohols, some acids and other chemicals.

CAB has similar properties to acetate, with improved dimensional stability, heat resistance, toughness and weatherability. It also has a noticeable odour.

CAP has properties similar to those of CAB but is slightly harder, stronger and stiffer, and has better heat-resistance. Light fittings, knobs, toilet seats, decorative trims, etc. have been produced from cellulose materials.

Chlorinated polyether

This material is used in corrosion-resistant applications. It can be moulded, extruded or applied as a coating. Extruded pipes and pipe fittings and protective linings are common uses.

Fluorocarbons

The most important fluorocarbon is polytetrafluoroethylene (PTFE) which has outstanding chemical resistance and electrical insulation properties. It is very tough and has the lowest coefficient of friction of the plastics. Teflon belongs to this group of materials.

Special processing techniques are required for this material, but some building applications include specialised heat-resisting pipes and fittings, insulating tapes and sheets, pipe thread seals.

Two other fluorocarbons are used which are more easily processed - PCTFE and FEP - but these do not have the same high qualities as PTFE.

CHEMICALLY BASED PRODUCTS

Nylons (Xylon)

Nylon in building is usually confined to bearings, etc. of doors and hinges, and to carpets. Its low friction, toughness and hardness are benefits, but it does absorb water, which can cause swelling.

Nylon carpets have a tendency to produce static electricity when dry which can cause shocks to occupants of the carpeted areas.

Polyethylene

These materials are tough, flexible, water impermeable, and have good chemical resistance. However, they suffer from low tensile strength and have poor weathering qualities. They are used mainly as moisture barriers, dampcourses, pipes, electrical insulation.

Polycarbonate (Lexan, Merlon)

This rigid and dimensionally stable material is transparent and has excellent impact strength but is affected by solvents and scratching. It is used for safety glass and lighting situations and is especially useful as a vandal resistant material.

Polystyrene

Easily fabricated, this material is cheap and has been used widely. Hard, rigid and transparent, it is brittle, with low heat resistance and of poor performance when exposed to weather. Foamed, it can be used for insulation.

Polyvinylchloride (PVC)

This material is made in two forms, plasticised and unplasticised.

Plasticised PVC contains up to 50% of this component and is then flexible, cheap and self-extinguishing. It is used for floor coverings, electrical insulation and coated fabrics. It becomes stiff at low temperatures and suffers from plasticiser migration.

Unplasticised PVC is tough, abrasion resistant, chemical and water resistant, weather resistant when stabilised, selfextinguishing and low in cost. It is used for pipes, conduits, window frames, panels, siding, synthetic wood and bricks.

Polymethylmethacrylate (Acrylic, Perspex, Lucite)

These products are transparent, with excellent optical properties, dimensional stability and excellent sun and weather resistance. They have low scratch resistance and limited chemical resistance, and are used for glazing, lighting fixtures, domes, etc.

Polypropylene is used in furniture and pipe fittings and is particularly resistant to destructive agents such as organic solvents, acids and alkalis. It can be welded to produce excellent pipe joints.

THERMOSETS

In these plastics, the long molecular chains are cross-linked to give three-dimensional networks so that the chains will not slip independently, even when heated.

Some thermosetting materials in use today are outlined here.

Epoxides

These materials have good strength and stiffness, low shrinkage, low moisture absorption and good resistance to acids, alkalis and solvents. They can be used as glues or jointing compounds, as floor surfacing and repair materials, in paints and surface coatings, and as decorative panels.

The epoxy glues and surface coatings in two-pack systems are now well known and highly regarded for their efficiency and reliability. As clear coatings they surpass all varnishes for toughness and durability. The main disadvantage is the production cost of the epoxies.

Phenolics (Phenolformaldehyde, PF)

These plastics are stiff and strong but brittle and dark in colour. They are good electrical insulators up to 150'C. Strong alkalis and oxidising agents will destroy them.

Mouldings and laminates are made from phenolics. Paper, cotton, nylon, and glass fibres are often used as reinforcements.

Electrical goods, toilet seats, surface coatings, waterproof adhesives are all building uses, in addition to the core stock for decorative laminates.

Urea formaldehyde (U-F)

U-F is hard and abrasion resistant, with good chemical resistance. Strong mineral acids will attack it, and moisture absorption and shrinkage are defects.

Knobs, handles, adhesives for plywood and particleboard are uses already established.

Large vaults can now incorporate curved plastic sheeting which can be clear or translucent.

Polycarbonate sheets attached to stainless steel tubing to form sunscreens.

CHEMICALLY BASED PRODUCTS

BUILDING PRODUCTS MADE FROM PLASTICS

Adhesives

Some important advances in building technology have been made possible by the strong weather-resistant adhesives produced from plastics. Timber glues particularly have enabled large and very strong members of plywood or laminated timber to be used structurally and in exposed locations.

Contact adhesives, epoxy glues, etc. can now be used to bond almost any two materials together with strength and life previously impossible. See Lysaght Referee for adhesives for joining steel sheet. Sheet glues that are heat sensitive can be ironed or hotpressed between components to provide clean and speedy gluing processes.

Carpets

Nylon carpet readily generates static electricity when the relative humidity falls below 65%. The charge occurs mainly under the action of footsteps. The accumulation of charge is lessened by the use of anti-static nylon which has stainless steel threads in the pile. For critical locations, however, such as computer rooms and in aircraft, larger diameter metallic wires are more effective and more durable.

In regard to fire and plastics in carpet, the Experimental Building Station has this to say - 'Cotton, viscose, nylon, acrylic and polyester will spread flame fairly readily ... Plastics used for fibres and rubbers used for backings can be formulated in various ways and they can incorporate fire retardants. These factors, together with the carpet construction and fibre form, can affect the flammability of particular carpets.' (Refer NSB 127.)

Cellular plastics

Many plastics can be prepared with a cellular internal structure containing air or gas in the cells. The cellular structure may be discrete or interconnecting.

One method is to incorporate a foaming agent which releases gas when the base polymer is heated to a certain temperature. Another technique incorporates chemicals which produce the polymer and a gas at the same time.

Cellular plastics are generally very light of weight and extremely good insulators. The base materials are usually polystyrene or polyurethane.

Foam sandwiches are made faced with solid skins of the same or a different base polymer. These give exceptional stiffness for unit weight.

These products are often used also as a core between casings of metal, concrete, etc.

Concrete formwork is another growing application in which complex shapes can readily be produced and minor on-site modifications effected.

Gaskets and weatherstripping

Standardised shapes of PVC and elastomers are now produced to serve as gaskets and weatherstrips. These have opened up new methods of waterproofing and constructing such items as joints in concrete walls, glazing in metal and concrete frames.

Silicones

Silicones, which are now widely used for weatherproofing joints between materials, consist of silicon, carbon, hydrogen and oxygen. They are chemically inert, being affected only by strong alkalis and hydrofluoric acid. Silicone oils show little change of viscosity with temperature and do not oxidise or harden. Silicone solids use inorganic fillers and can be cured or vulcanised to a rubbery product. Silicone gaskets as jointing components for glazing have been widely and successfully used.

Laminates

Laminates were the first major plastics materials to make a visible impact on the building industry during the 1950s. They use the well-proven technique of building up a large, stable and tough sheet material from thin layers of varying properties.

Both thermoplastics and thermosets are used. They are pressed and heated between flat parallel plates, causing the plastics to flux and form a uniform matrix. The core plies can be of the dark, strong plastics, while surface layers are suited to the finished colours, patterns or textures.

The basic material is frequently melamine formaldehyde, which is a thermoset.

The finished products are dimensionally stable, hard, abrasion resistant, non-absorbent and easy to clean, but they are attacked by some chemicals.

The well known decorative sheet materials Laminex & Formica, and others, are examples. As surface finishes glued on to fibre boards they have been very widely used and come in horizontal & vertical grades. More versatile versions of these materials are now available as Laboratory; Compact (double sided); Wet area; Fire retardent; Graffiti resistant; X Ray resistant; Cubicle partitions, and other grades. The manufacturers should be contacted for detailed information.

Metal protection (Colorbond, etc.)

Numerous products are now made from light steel wire or sheet with a coating of plastic for corrosion prevention. The plastic used is often vinyl, which is flexible and adheres well to the metal. It is made in a range of colours. These products are used in many metal roofing situations, for skirtings, pipe covers, airconditioning equipment, furniture and accessories.

Steel window frames have also been given much improved performance by plastic coating techniques.

Pneumatic structures

Sheet plastics have been exploited to develop air-supported structures such as domes, vaults and inflated ribs, pillows, roofs, etc. where the lightness and translucency of thin sheets are often advantages. Heat welding is often used to make up special large shapes from standard sheet sizes.

Rainwater goods

Roof gutters and downpipes are now available in PVC materials which should give long life without corrosion and are easily installed without specialised equipment or skills. They can be readily cut with a hacksaw. Surface drains and grates are now being made from recycled plastics.

Reinforced plastics

Many plastic products incorporate reinforcing materials in the form of fibres, sheets, strands, etc. Their purpose is to improve the mechanical performance and stability of the end product.

The commonest materials used for this purpose are glass, asbestos and synthetic fibres. The production methods vary from crude hand packing to sophisticated spinning and moulding aimed at reproducing many copies of identical products.

For best performance and appearance the fibres should not be visible or felt on the surface. To ensure this a finishing 'gel' coat of pure plastic is applied to many products.

Building panels incorporating windows have been made up in factories and bolted in place on site using reinforced plastics. See also GRP under 'Glass Products'.

A striking development with these plastics has been in tension membrane structures, usually in the form of roofs and sails. These are now made from tough Polyvinyliden Fluoride (PVDF) which resists wear and pollution, and provides a long life.

Roofing materials

Many roofing components have been made to match or fit with standardised systems of metal and other materials for roofs. This applies particularly to skylight-type items where the breakage risks are minimal compared with their glass forebears.

Special shapes can be more readily made than with most other materials. Ribbed, folded, domed, corrugated shapes have been produced for roofs of small and large dimensions.

Virtually unbreakable and translucent materials are available in polycarbonates.

Vapour barriers and waterproofing

The large widths available in sheet plastics have made these products popular as moisture barriers in many locations such as waterproofing under concrete slabs on the ground.

These polyethylene sheets can be sealed at joints with tapes or brush-on adhesives and are of reasonably heavy gauge, coloured and branded for clear identification.

Many damp-proof courses and flashings in masonry are now made of plastic strip materials which are usually textured to improve the rigidity and grip between the mortar and the plastic.

Waste piping and venting

Standardised pipes to meet sewerage authority standards, and for waste water and agricultural drainage are now made from plastics. Because of the smooth internal surfaces, long pipe lengths with few sealed joints, these plastics have a very good performance record and seldom become blocked.

Many varieties and components are available and include appropriate elbows, junctions, traps and related fittings from small diameters of approximately 25 mm to 150 mm, and in various wall thicknesses.

CHEMICALLY BASED PRODUCTS

The materials used for these pipes and fittings are ABS, PVC, PE, PP, giving light, tough, corrosion-resistant products that can be readily fabricated using adhesives at joints.

Piping exposed to hot weather or sunlight may need extra supports and consideration regarding expansion or freezing.

Water supply piping

Corrosion resistance and low thermal conductivity are two reasons why plastic water pipes have been approved by some authorities for use in buildings.

As temperatures and pressures can be high in hot water services, special co-polymers are needed in order to be satisfactory. Added supports may also be needed to avoid lines sagging when hot. Expansion and contraction effects also must be allowed for.

Accessories

Many uses for plastic products have been developed such as adhesives for special situations; chemical anchorages for masonry & concrete; flexible crack fillers; injected damp courses etc. and plastics are used as parts of many hardware items.

Typical tension membrane roofs made possible by modern plastics technology and computer modelling.

A demonstration assembly of plastic piping showing some of the many components available for waste water.

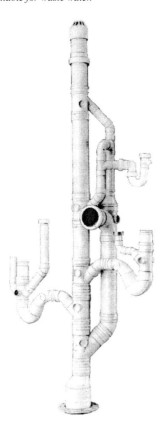

SUMMARY OF CHEMICALLY BASED PRODUCTS

The variety of chemical composition and the complexity of these products are such that they cannot be summarised in the specific manner attempted for other types of materials. Each product needs to be checked out for its particular application.

Structural strength

Materials such as nylon are available with very high tensile strengths yet are flexible and able to recover to original dimensions (high elasticity).

High compressive strength is also available, especially in some thermosetting products and epoxies, but such components are usually of a small detail nature rather than for walling or basic structure.

Manufacture and forms available

Manufacturing methods are widely varied and frequently complex, being carried out in rigidly controlled conditions of temperature, pressure, etc. However, end products are frequently soft enough to be site worked to a limited degree.

The variety and range of available products is impossible to summarise.

Water and freeze/thaw effects

Many of these products are formulated deliberately to resist water penetration and its effects. New applications of a plastic should be specified only after accredited testing laboratories have given satisfactory reports.

Low temperatures cause brittleness in some plastics.

Temperature effects

Climatic temperatures affect the flexibility of many chemical products so that local conditions need special attention regarding critical temperatures for specific products.

Where mastics and caulking are concerned their function is often to contract when other components expand, and vice versa, without losing adhesion or effectiveness.

Some products such as polyurethane foam are extremely good thermal insulators.

Site glueing or welding is often possible with heat applied to thermoplastics.

Fire resistance, flammability, emission of smoke or toxic fumes in fire are common defects in chemical products and are now receiving attention. Study and performance reports should be available from testing authorities or manufacturers.

Ultra-violet radiation

This is one of the hidden weaknesses of many chemical products. Asphaltic and bituminous materials are broken down over a few years of exposure. Many products are scarcely old enough to have revealed similar weakness, so all should be treated carefully in critical exposed situations, and protected from direct sunlight where practicable to achieve maximum performance.

Electrolytic or other special effects

Many of the vinyl-type products in buildings are used specifically for electrical insulation. Some products can become electrically charged through climatic conditions, leading to unexpected effects such as the nylon carpet causing shocks to a building's occupants. Chemical reactions with other components can occur, especially between chemical products.

The toxicity of plastics has resulted in numerous products being banned from factories and places of public assembly because of their danger in fire. This applies particularly to those incorporating chlorine or cyanide in their manufacture.

Creep under load and degradation by ageing are other aspects requiring careful consideration, and these are not well tested and documented so far.

Acoustic qualities

Acoustic qualities vary greatly between products, dependent on density, porosity, surface finishes, etc. Some products are used as sound absorbers in acoustic ceilings and wall panels, these usually being light, open-pored materials, which can be very efficient.

Mastics have an important role in sealing acoustic leaks through air gaps in windows, door openings, etc.

Flexible brush-type materials and felts are also produced which assist in sealing movable units such as doors and window sashes.

ACKNOWLEDGEMENTS
Thanks are due to the following people who contributed material and information for Chemically Based Products and Plastics.
John Hartley - Vessey Chemicals
Dr. Gordon Renwick - NSWIT Department of Materials Science
Ross, Henry, Bill, Marty and Peter - Students NSWIT

References
The following references are recommended for further study:
DIETZ, A. G. H. - Plastics for Architects & Builders, MIT Press, 1969.
OWENS CORNING FIBERGLAS - FRP. An introduction to Fiberglas Reinforced Plastics/Composites. Fibreglass Limited - The spray up moulding process.
COMMONWEALTH SCHOOLS COMMISSION. Comparative Suitability of Materials &Finishes for Schools in Australia, 1982.
BASF Publication - EPS Insulation.

Standards
ASCA55:1970	*Code of recommended practice for bituminous fabric roofing.*
AS 1160-1988	*Bitumen emulsions for pavements*
AS 1366	*Rigid cellular plastic sheets for insulation*
AS 1507-1980	*Road tars for pavements*
AS 2008-1980	*Residual bitumen for pavements*
AS 2150-1995	*Hot mix asphalt*
AS 2157-1980	*Cutback bitumen*
AS 1604-1993	*Preservative treated timber*
AS 1607	*Water repellants for timber & joinery*
AS 1627	*Preparation of metal surfaces prior to painting*
AS1884- 1985	*Resilient sheet & tile floor coverings*
AS 2032-1977	*Installation of UPVC pipe systems*
AS 2033-1980	*Installation of Polyethelyne pipe systems*
AS 2131-1987	*Contact Adhesives for thermoset laminates*
AS 2150-1995	*Hot mix asphalt*
AS 2157-1980	*Cut back bitumen*
AS 2310-1980	*Glossary of paint & painting terms*
AS 2311-1992	*The painting of buildings*
AS 2376 .1 & 2	*Plastics building sheets - PVC & GRP*
AS 2424-1991	*Plastic building sheets for roofing systems*
AS 2734-1984	*Asphalt (hot mix) paving*
AS 2754.2-1991	*Polymer emulsion adhesives*
AS 2921-1987	*UPVC, wall cladding*
AS/NZS 2310-1995	*Glossary Of Paint & Painting Terms*
AS/NZS 4200	*Pliable building membranes & underlays*
AS/NZS 4256	*Plastic roof & wall cladding materials*

Numerous other Standards also refer to plastic pipes, for water, sewer and gas.

CSIRO NSB's

39 Damp proof courses and flashings
73 Epoxy resins
131 Bituminous driveways
137 Fire hazard of furniture &furnishings

Large fibreglass panels used as cladding and window frames in a city office building illustrate the growing intrusion of plastics into the building industry.

INCOMPATIBILITY OF
MATERIALS

Metallic Corrosion, Stains; Surface finishes, Chemical reactions;
Differential movements; Design detailing; Cleaning

In the foregoing chapters efforts have been made to identify the peculiar properties of materials which may lead to problems in building situations. The range of situations where differing materials are now juxtaposed in buildings is limitless and changing. It is very difficult to predict many of the problems which will occur. When only a few basic materials were used on the exterior of buildings, the tradesmen, builders and architects were usually aware of troublesome combinations and from craft knowledge avoided most serious problems. These tended to be largely the electrolytic actions of metals, and salts in masonry.

The large-scale intrusion of the chemically based materials, the widespread pollution and use of new construction methods have led to new problems of both chemical and physical nature in buildings.

Incompatibility of building materials can be categorised to some extent.

The following are suggested as appropriate groupings.

A. Corrosion of metals.
B. Stains and discolouring effects.
C. Problems with surface finishes.
D. Chemical reactions between materials.
E. Differential movements of materials in contact.
F. Defective design detailing.
G. Cleaning problems.

The following notes try to illustrate known cases of building material problems arising from incompatibility under these headings.

A. METALLIC CORROSION

Corrosion in its destructive role in building can be defined as `the undesirable deterioration of construction materials by electrochemical action resulting in loss of functional or aesthetic value'.

Galvanic reactions

The galvanic scale has been explained under `Metals' and problems under this category can largely be anticipated using that knowledge. However, many `accidents' can occur, especially when work is carried out on roofs after completion of the main building contract. Offcuts or filings of metals left where water comes in contact with them have caused rapid corrosion.

Zincalume is particularly subject to this problem, due to the zinc and aluminium in it being at the more active (anodic) end of the galvanic scale, which means they can be readily corroded if in contact with metals toward the passive (cathodic) end.

The use of lead in association with zinc or aluminium will promote corrosion, so flashings need careful selection for metal roofs. Steel nails or screws should not be used with aluminium or zinc roofing, unless coated with zinc or cadmium.

Copper goods should not come into contact with, or drain onto, zinc, aluminium, zincalume or galvanised materials. Clips and fixings should be of similar metallic bases.

Lead-based or graphitic paints should not be used on aluminium.

GUIDE CHART

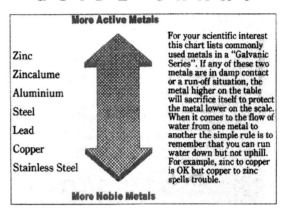

Water may safely flow downwards between metals in the electrochemical series above but not upwards. Reproduced from Building with Australian Steel,' by courtesy of BHP

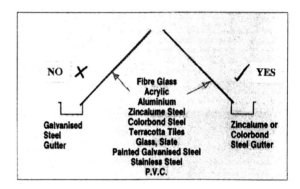

Roof & gutter incompatability diagram, by courtesy of BHP.

Water-metal corrosion

Most ferrous metals rust when in contact with air and water. This is particularly apparent in warm humid atmospheres when the alloys used for pipes and structural steels are most reactive.

For steel to be well protected against severe industrial, or marine environments within 300 metres of the sea, galvanising plus another surface coating is advisable. Sheet steel protection is best provided by galvanising or zinc aluminium coating followed by a PVC plastic coating applied in the factory. Any raw edges should be sealed on site.

The Lysaght research

The following late 20th Century research was undertaken by the John Lysaght Research and Technology Centre based at Port Kembla.

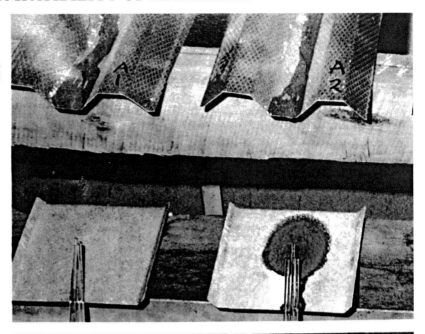

Apparatus was set up to establish the relative performance of ZINCALUME* steel and galvanized steel under different roof catchment conditions.

Distilled water containing 10 mg of chloride (salt) per litre dripped continuously onto pairs of roofing sections. Each pair has a different surface

The photographs on this page were taken after two years' testing.

1	Aluminium to ZINCALUME gutter (left) Aluminium to galvanized gutter (right)
2	COLORBOND to ZINCALUME gutter (left) COLORBOND to galvanized gutter (right)
3	ZINCALUME to ZINCALUME gutter (left) ZINCALUME to galvanized gutter (right)
4	Galvanized to ZINCALUME gutter (left) Galvanized to galvanized gutter (right)

4

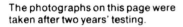

Research evidence when available is very useful in avoiding problems of incompatibility of materials

INCOMPATIBILITY OF MATERIALS

The John Lysaght Research & Technology Centre has shown galvanised steel to be susceptible to rust from water drips containing salt, particularly when coming from a roof of dissimilar material, of either metallic or inert surface. See illustrations, and refer to Lysaght's Technical Bulletin SC 1-1 June 1983 for detailed recommendations.

Galvanised piping is often used for water supply and gives reasonably good life where the water is 'hard' or not acidic. Corrosion in such systems occurs if carrying low to moderately hard water at junctions with brass or other copper alloys. Plastic connectors can be used to eliminate this. Internal corrosion can greatly reduce the effective pipe diameter.

In hot water systems, mixtures of differing metals for pipes should be avoided, as the corrosive effect tends to accelerate at higher temperatures. Copper and brass are permanently resistant to water. After an initial scale is formed in water pipes, further action ceases. Some rare cases of breakdown in copper pipes have been traced to extremely 'soft' water, i.e. no alkaline content and no acid.

Aluminium in a coastal atmosphere acquires an encrustation from alkaline attack. In less corrosive airs, a thin aluminium oxide is formed which protects the metal from further action.

Alkali-laden materials containing mortar, cement or concrete should not be allowed to splash aluminium, as this will cause surface etching unless quickly removed.

Industrial atmospheres are commonly acidic and therefore corrosive on all metals. This acidity is increased considerably where ponding occurs in gutters, etc. Non-metallic coatings are sometimes advisable in such situations.

Fixings

All fixings need to be of materials compatible with the metal secured. Manufacturers recommendations should be respected or advice sought for unusual situations.

References

NSB 79 Corrosion of metals in building
Lysaghts Technical Bulletin SC-1.
.NCRB Problem Areas in Building. Rust and Corrosion.

B. STAINS

Stains and colour effects from surface drips should generally be avoided by good design detailing, but problems can arise due to poor detailing or conditions during progress of the work. Reference should be made to NSB 59 - 'Cleaning Brickwork', which outlines methods of treatment for various types of staining.

Copper stains

One of the common problems arises from water dripping off copper causing greenish stains on materials below. Where the materials are porous this can be very difficult to remove and becomes a stubborn stain, even on surfaces such as porcelain enamel. Copper water pipes can cause basins and baths to mark badly, if a leaking tap is not fixed promptly. Copper rainwater goods can cause similar problems.

Rust stains

Water from exposed steel or ironwork will carry rust stains to surfaces below. These will penetrate porous materials and become impossible to remove from masonry.

Timber stains

Many Australian eucalypt timbers produce a dark brown stain when wet and this can affect masonry below it. This can be a serious problem if face brickwork, concrete or stonework is affected during construction.

Western red cedar timber contains a chemical which reacts with steel so that nails used in this material need to be copper, monel metal or hot dip galvanised. Steel nails will produce a bad black stain on the surface. Aluminium is also reactive with W.R.C. in a similar manner if unprotected surfaces make contact.

Efflorescence

Efflorescence on brick or stonework is caused by migration of salts through the porous elements. The salts can be from the ground, sand, cement, bricks or stone and proceed usually as far as rising dampness extends. The salts appear on the masonry in the form of a whitish surface stain and may be removed, but will often reappear until the offending source of salt is exhausted. For this reason, any material containing salt is incompatible with masonry construction. Sand should be pre-washed if from seaside sandhills. Rising dampness should not extend above dampproof courses where provided, but in retaining walls etc. moisture from behind the wall causes staining on the face.

C. PROBLEMS WITH SURFACE FINISHES

Lack of adhesion

In applying finishes to basic materials, problems frequently arise because the sub-surface does not suit the selected finish. These relate usually to the texture of the base surface. Some materials, particularly the cement-based ones, adhere best to a rough surface and may not adhere adequately to a smooth one. Porous cement, concrete, brickwork, stonework are suitable bases for further cement or lime-based mortars, plasters and renders. Attempts to apply these to very smooth hard products of the same masonry types can cause lack of adhesion and failure.

Lack of adhesion by applied cement render or topping to concrete can also be caused by the use of unsuitable curing compounds such as chlorinated rubber. More suitable in such situations are compounds based on PVA which are progressively degradable.

Waterproofing membranes

Concrete roofs, which need to be waterproofed by the use of a membrane, should be well cured. A useful estimate of time required for natural curing is 25 mm per month (e.g. a 150 mm slab will require 6 months to dry naturally). As it is not always practicable to delay placing the membrane, one capable of allowing the vapour from the slab to dissipate must be chosen, e.g. Scotchclad, hypalon, isobutelyne. An impermeable membrane will cause bubbling as the vapour escapes from the curing slab.

Special tiling grouts

Acid-resisting grouts used with floor tiles in certain situations need to be very carefully handled to avoid spreading over the tiled surface, as these cannot be removed effectively by conventional acid washing methods.

Painting galvanised surfaces

The problems experienced with applying paint to galvanised or zincalumed steel surfaces are closely related to the smoothness of the metal. These coatings are particularly smooth and the paint's adhesion to the metal surface will last only a few months. For preparations recommended in this situation see chapter on Paints.

Silicone sealants and paint

Silicone sealants will not accept paint coatings. This can lead to problems if areas adjacent the silicone seals are smeared and not carefully cleaned with suitable silicone solvent.

Acrylic paints in contact

Acrylic painted surfaces are thermoplastic and should not be brought together in situations such as doors and window frames and sashes. The paint coats weld together and effectively prohibit the opening of such elements.

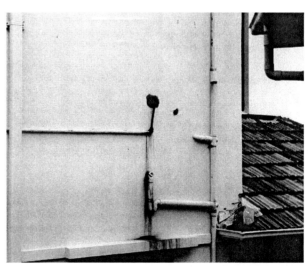

Rust stain due to drips from above onto steel waste pipe on wall in seaside atmosphere.

Primers and undercoats

All paint applications should be applied using primers and undercoats strictly in accord with manufacturers' instructions, as the chemicals involved in modern paints can interact and cause breakdowns if not used correctly.

With paints of the old high gloss variety, it is undesirable to apply one gloss coat immediately over another without roughening the surface with fine sandpaper. This helps to provide a grip or key for the finishing coat and minimise the risks of the paint flowing downwards causing 'curtaining' effects.

Bases for gloss finishes

With pre-finished materials such as plastic laminates, glossy surfaced flexible floor coverings, applied with thin flexible rubbery adhesives, the surface must be smooth, free of visible bumps or hollows. The gloss surfacing exaggerates any defects, especially when seen looking toward a light source at or near the surface level. Plain unpatterned materials more readily produce this visual defect than patterned or textured ones.

Factory floors

Factory and warehouse floors of concrete can be quickly dusted up by some types of heavily loaded trolleys. Applied concrete toppings will seldom stand up to this traffic so the surface needs to be finished integrally with the slab. Similarly, tiled pavings will break under wheeled traffic.

To decide an appropriate pavement surface, an on-site mockup may be created to compare results under simulated conditions to give an accelerated testing system.

See also Chemical Reactions caused by milk, etc.

D. CHEMICAL REACTIONS BETWEEN MATERIALS

Materials to be stored or used in a building can sometimes be incompatible with building materials.

Salt warehouses, for instance, generate highly corrosive conditions for ferrous metals and these should be avoided. Timber and cement-based components are usually a better choice in such situations.

Milk contains lactic acid, which is very destructive on cementbased materials. Concrete floors in dairies, milk factories, etc. need special pavements and surface treatments because of this.

Ammonia-bearing materials, such as used in some adhesives, can be damaging to copper and brass, so copper piping or brass fittings should be avoided in ammonia-laden atmospheres.

Enclosed swimming pools generate warm moist atmospheres, which very quickly attack any mild steel which is inadequately protected. Hot dip galvanising is the best protection, with use of stainless steel for fittings and equipment wherever possible.

Chloride-bearing concrete or brickwork will cause all metals in contact to corrode. Some concrete slabs in which calcium chloride has been used as a crack-inhibiting agent have developed serious problems with reinforcing rods. Hydrochloric acid drips used as a cleaning agent can also cause problems.

Galvanised steel and lead will corrode in wet conditions where in contact with cement mortar or concrete. Separation by wrapping or placing a plastic sleeve over pipes is often adequate protection.

Magnesite floor surfacing consists of calcined (or burnt) magnesite and organic and inorganic fillers mixed with a solution of magnesium chloride. It is trowelled onto the structural concrete or timber base floor to give a smooth surface.

Magnesite is corrosive to metal objects in direct contact with it. Steel or aluminium door frames, mat wells, pipes, conduits and the like should be insulated from it, by bituminous coating, chemical-resistant tapes, plastic or hardwood strips or 25 mm of dense mortar or concrete. Sheet vinyl or linoleum coverings will usually give adequate protection to stove, refrigerator or dishwasher legs, where the sub-surface is magnesite. Magnesite should not be used in locations such as laundries subject to water as acids can develop and permeate concrete sub-floors, corroding reinforcement.

For detailed description see NSB 117 - Magnesite Flooring.

Treated Timber Fixings.

While Galvanized fixings are commonly used for treated timbers, cases of reaction have occurred, and local checks are advisable.

Salts in masonry or ceramic products

Porous products of geological origin, such as stone, brick, terra cotta and concrete, can be severely affected by salt penetration. The salts are carried into these materials by water, often transported by capillary attraction. Obviously the more densely compacted the material, the less likely it is to encourage capillary migration of salts.

The salts may come from ground water, the atmosphere, the raw materials, the production process or other building materials. Salts most commonly causing building problems are sulphates of sodium and magnesium, or sodium chloride, which can be present in significant quantities in coastal atmospheres.

Magnesium sulphate is the most damaging of the soluble salts and can cause serious trouble.

Sodium sulphate is the cause of surface efflorescence on brickwork and masonry, which can be removed but often returns until the source is eliminated.

Calcium sulphate attacks the cement content of mortars and plasters.

Sodium chloride (common sea salt) is most active in attacking poorly burnt clay products or those fired at low temperatures. It can cause disintegration of surface material.

Glazing or surface coatings are seldom 100% effective in protecting ceramic products from surface deposits, as water can enter via very small crazes in the glaze and cause a build-up of crystallised salts as evaporation occurs.

Strong cement mortars or renders can contribute to salt problems by adding to the total salt content of the building and affecting materials via migratory action.

Proper design and construction of dampcourses and careful choice of bricks or stones in areas subject to damp, especially retaining walls, parapets, etc., will minimise problems.

In coastal areas affected by sea spray etc., modern cement roofing tiles are usually more reliable than terra cotta tiles, which can show serious salt damage within a few years of installation. Advice should be sought from the Clay Brick & Paver Institute Wentworthville 2145 for difficult conditions. They have numerous publications which may be helpful. See also Brickwork chapter.

A 20-year-old fence of porous bricks badly affected by salty marine atmosphere.

Hard smooth bricks have resisted salt attack but coloured mortar joints have been eroded.

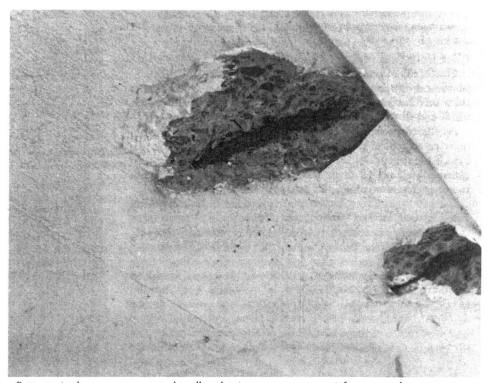

Porous or inadequate concrete cover has allowed moisture to penetrate to reinforcement and generate rust which has `blown' the cover completely.

E. DIFFERENTIAL MOVEMENTS OF MATERIALS

Materials which are bonded to one another by cement mortar or other inflexible means can be a source of problems in building, especially when the surface material is exposed to significant temperature variations, or when the two materials display differing behaviour patterns over time.

Surface veneers such as ceramic tiling exposed to weather and sunlight will be subject to daily thermal expansion and contraction and possibly to ceramic growth. The veneer protects the base material from much of these effects and the adhesive is obviously the location at which tension will develop because of these differential movements. As cement mortars are inflexible, they are weak in tension and cannot withstand such continuous stress variations. Cracking develops and adhesion is lost so the tiles become drummy and can readily fall off.

This effect can be minimised by fixing external tiling in relatively small panels, not exceeding 3 to 4 metres maximum dimension and allowing expansion joints filled with a flexible material, such as polysulphide or polyurethane, between the tiling panels or any adjacent wall or projection.

By using a flexible rubber- or latex-based adhesive which can withstand external conditions, the problem can be minimised, but expansion joints between panels are still advisable, both externally and internally.

This problem is exaggerated when the structural support for the tiles is a cement or concrete product, which continues to shrink as it ages, while the ceramic products tend to expand. Refer NSB 124 - Internal Ceramic Tiling.

The flexible adhesives are also more effective in resisting fracture due to vibration, which is often an important consideration when buildings are close to railway lines or major roads.

Facings on prestressed concrete

The built-in compressive stresses in this material ensure it will expand very little under stress and will be maintained in its most compact longitudinal dimension. Its initial shrinkage is greater than normal reinforced concrete. Any facings therefore should be provided with flexible contraction joints and not be fully grouted up or rigidly fixed. Engineering advice should be sought regarding such applications.

Parapets to concrete roofs

Where a concrete flat roof requires an edge parapet, it is advisable to avoid making a change of material.

Where brick parapets are used, the tendency for the concrete to shrink and the brickwork to grow leads to noticeable differential movements, distortion of membranes and flashings at edges and cracking.

Brick infill in concrete frames

One of the most serious and troublesome results of differential movement of materials in Australia has been the displacement of brick infill panels used as external walling between concrete columns and floor beams in multi-storey buildings.

As the concrete shrinks and creeps with age, the dimensions between structural components diminish. At the same time, the brickwork dimensions are growing. The resulting excessive compression has buckled single-skin brickwork, creating a danger of masonry elements falling, and cracks for weather penetration. See NSBs 134 and 135 on Differential movement in buildings clad with clay bricks.

The cure lies in using only matured bricks, mortar containing lime and a minimal cement content (6:1:1 sand, lime, cement) and building expansion gaps into each panel to absorb these movements. This is desirable at the top of each panel as well as ends.

Refer also to NCRB 1986 Problem Areas in Building.

Different vertical supports

In multi-storey buildings, using different structural materials, systems or loadings adjacent to each other can be troublesome. For example, some steel-framed multi-storey buildings are constructed using a slip form reinforced concrete lift core. The core construction usually precedes the surrounding frame erection, but allowance has to be made for the progressive reduction in height of the core as the creep in the concrete accumulates under the effects of the increasing load. Engineers can usually calculate this closely, so that the final floors resting on both the steel and concrete supports will still be tolerably level.

Steel on the contrary expands with heat, and this can cause some differences in movement between externally exposed and internally protected members over considerable lengths.

Highly stressed perimeter columns sometimes contract more than lowly stressed supports and this also needs calculated consideration.

F. DEFECTIVE DESIGN DETAILING

Concrete and glass exteriors

Details produced in the design office frequently contribute to building problems. A common fault relates to the disposal of surface water due to rain. The choice of materials greatly affects these details.

For instance, if constant dripping from concrete or cement mortar causes splashing onto glass, the glass surface can become alkaline etched and lose its transparency. A stone surface or lime mortar would not produce the same result.

The whole field of detailing sills, cappings, etc., where concrete elements are involved needs much more case study evidence to produce recognised effective methods such as were commonly illustrated and understood for brick and stone when these were the standard materials. It is desirable to produce an even washing of rainwater over the whole facade surface in storm conditions, otherwise heavy staining develops, marking the areas of concentrated flow.

Ferrous metals and masonry

Most ferrous metals (except the stainless steels) corrode noticeably in atmospheric conditions. Where they are built into any form of masonry this eventually leads to rust formation, which expands the space occupied by the metal and causes cracking of the masonry.

This effect can be clearly seen in iron and steel railings built into stone, brick or concrete and in arch bars or lintels supporting brickwork over openings. While the damage may take a long time to develop in dry unpolluted atmosphere, it can become apparent within a few years in warm coastal conditions with salt-laden air.

The salt in the coastal atmosphere penetrates into the morter in which wall ties are located in cavity brickwork causing galvanised ties to rust and break.

Stainless steel or non ferrous metal ties should be used in such locations.

Modern galvanising, zinc paints and epoxies should be used to protect the base steel from these defects for any permanent structure, before the metal is built in.

On the special weathering steels, such as 'Austen', the initial development of the oxide coating unfortunately still produces sufficient drip stains to be troublesome on materials below.

Timber and moisture

External timbers should not come into contact with concrete floor slabs or the ground so that moisture can be trapped in the junction between the two. A galvanised steel shoe embedded into the concrete and elevated approximately 75 mm above the base is a useful device to minimise this problem. If the upstanding leg of the shoe is recessed flush to the timber post, this helps to minimise water entrapment at this junction.

Ineffective junctions

The juxtaposition of certain materials sometimes makes it extremely difficult to produce an effective detail, especially when waterproofing is a consideration. This applies particularly to the joint of a waterproof membrane upturn with a concrete wall. Unless a properly fitted reglet is cast into the concrete, it is almost impossible to achieve satisfactory weatherproofing as can usually be obtained with flashings built into masonry joints.

A similar problem occurs if one end of a urinal stall in stainless steel abuts a precast partition slab.

Such problems are much more the result of ineffective detail thinking than of incompatibility of materials. Nevertheless, they are commonly found in buildings and result from a lack of threedimensional thinking and the workability of the materials on the building site.

Cracking patterns in brickwork caused by rusting and expansion of unprotected steel lintels over windows.

Roof and ground water

Water normally moves downwards under the effects of gravity. However, situations can arise in buildings where a pipeline becomes inadequate or blocked, causing a quantity of water to build up vertically creating a hydrostatic head. This can be serious if roof water is under pressure in a downpipe, especially when there are agricultural or surface drains discharging into the rainwater system without means to prevent backsurge in connecting pipes. Such one-pipe systems, incorporating roof water and basement or sub-soil drainage, need very careful consideration to avoid backflooding under extreme storm conditions. Separate systems are preferable for safety and efficiency. Reflux valves will seldom suit the very low flow rates associated with agricultural pipelines. Open pits at junctions between the two types of drainage can help if external to the building, but internal pits can cause flooding problems.

Refer to NSB No. 58 Concrete Floors on the Ground for Commercial Buildings, NSB No. 89 Sub-Surface Drainage and NSB No. 84 The Weatherproofing of Buildings regarding appropriate systems.

G. CLEANING PROBLEMS

Problems with cleaning vary greatly according to the type of premises, the surfaces to be cleaned and the material to be cleaned off.

With stone and brick buildings, the recommended treatment for normal atmospheric city grime is to soak the surface thoroughly with clean water over a period then lightly scrub the surface.

Many harsher treatments involving steam, high-pressure water jets, detergents and other chemicals have proved to be damaging to the masonry and sometimes generative of rapid deterioration thereafter.

For internal surfaces, where more frequent cleaning is usual, mild chemicals, abrasives, water, etc. are quite commonly used but in each case these should be compatible with the surface material. Chemical cleaners often contain abrasives and bleaches which can damage surface colours and finishes if used frequently.

For areas such as industrial and commercial kitchen and toilet floors, where maintenance is costly and often done by contract cleaners, efforts should be made to avoid having bases of fittings and equipment supported from the floor. A plinth with coved corners and vertical face tiled with floor tiles is a very useful device. Where floor support is needed, stainless steel is one of the preferred materials to minimise moisture corrosion. Chipboards should be kept well clear of areas involving moisture.

Laminated plastic bench tops and the like can be very badly affected by harsh abrasives. Soft chemical-type cleaners are usually preferred.

CONCLUSION

As the generalists and designers in building matters, architects and builders are usually well served and assisted by numerous consultants (e.g. structural, mechanical, hydraulics engineers). However, there are no consultants for the wide range of materials now used in buildings.

Many manufacturers and their associations have technical advisory services relating to their own field of interest. These should be consulted whenever in doubt. However, the architect or builder carries the professional and legal responsibility for materials chosen, specified or built into the building. They are usually the only ones able to perceive in advance or on the site the whole interplay of materials and conditions likely to arise in the finished work.

As over 37% of claims arising under indemnity insurance policies involve unsuitable choice or use of materials, it is clear there is much need for improvement in understanding the many materials now used in building.

The foregoing text and accompanying illustrations should help in this direction. If they succeed in improving the performance of the building industry in any significant degree, the effort will have been worthwhile.

COMPARATIVE TABULATIONS

Structural Properties
Thermal Properties: Expansion; Reflectance; Absorption; Conductivity; Resistance; Capacity;
General tabulation; Fire Resistance Rating; Fire Hazard Indices.
Acoustic Properties: Comparative weights; Sound reduction indices.

INTRODUCTION

Quantification of the physical properties of materials is usually required for engineering design to assess loadings, compressive and tensile strengths, expansion, etc. Engineers and specialist consultants are more familiar with some of these figures than architects and builders who usually think in comparative terms rather than figures.

Tables of these values are available in trade catalogues and handbooks, but rarely set out in a format which provides for easy comparison.

For most metals, available figures are considered accurate. With stones, timbers, soils, bricks and in situ concrete there can be considerable variation. For practical reasons, average figures are accepted and used for design purposes, and the following tabulations are generally derived from a variety of authoritative sources. Where more detailed or precise information is necessary, reference should be made to the particular trade associations or manufacturers, who usually have technical advisory services available.

Structural Properties

In many building situations, it is necessary to calculate loadings and the effects of loads on structural members. This requires various performance characteristics for each material to be known. The most important ones are shown in Table 1.

Reference should also be made to Standard Codes for Loadings -
AS 1170 - 1981
BS 648 - 1964 - Schedule of weights for building materials

TABLE 1

STRUCTURAL PROPERTIES OF MATERIALS

Material	Density Bulk Kg/m³	Ultimate Tensile Strength MPa	Max. Allowable Tensile Strength MPa	Ultimate Comp. Strength MPa	Max. Allowable Comp. Strength MPa	Elastic Modulus MPa	Coefficient of Thermal Expansion Strain/°C	Max. Likely Moisture Shrinkage %	
Aluminium									
99% Pure	2 650	75				70 000	24	Nil	
	2 700	300	150			70 000	23	Nil	
Brass (60 Cu:40zn)	8 400	400-550	50-200			100 000	21	Nil	
Cast Iron, gray	7 200	100-600				180 000	10	Nil	
Clay) low	1 900	0.020	0.050	20				-0.080)	
Bricks) medium	1 900	0.030	0.100	28				-0.080)	
) high	1 900	0.040	0.100	80				-0.080)	
Concrete) low	300-2 000	2	0.500	14	7	10 000	12	0.030	
) medium	2 200	3	0.700	25	12	15 000	12	0.040	
) high	2 400	6	1.000	50	24	28 000	12	0.050	
Copper, annealed	8 700	220-360	25-180			120 000		Nil	
Glass		15-150		100-300		70 000	9	Nil	
Lead	1 120	17				14 000	30	Nil	
Magnesium alloy	1 760	140-280				45 000		Nil	
Polyvinylchloride	1 350	60				25 000			
Polystyrene	1 120	95					70		
Polyethylene	900	9				13 000	170		
Polythene	900	7-40					150-200		
Stone, granite	2 240			100-200			11		
limestone	1 800-3 000			40-150			3		
marble				70			4		
sandstone				25-75			12		
slate				50-70					
Steel, alloy	8 100	1 700	1 100			200 000	10	Nil	
hardened	8 000	1 400				200 000	11	Nil	
piano wire		3 000				200 000	10	Nil	
stainless	8 100	960				200 000	11	Nil	
structural	7 900	480	160		160	200 000	12	Nil	
Timber	g-12% mc		g-12% mc		g-12% mc			Rad.	Tang.
Ash	1 100-670	35	11-14		3.5-13	12-15 000	4-5	7	14
Blackbutt	11 200-850	35	11-14		3.5-13	12-15 000	"		
Box, brush	1 200-900	40	14-17		5-16	15-18 000	"	2.5	6
Douglas Fir	650-560	55	8-10		2-10	10-12 000	"	2.5	4
Gum, grey	1 300-1050	55	17-20		5-16	16-21 000			

THERMAL PROPERTIES
Expansion
The simplest and best understood effect of heat on building materials is the expansion and contraction which take place with temperature variations.

Physicists speak of the coefficient of linear expansion of materials and have figures for each material. For building purposes, the following type of comparative tabulation probably gives more useful information and is readily understood.

TABLE 2
TEMPERATURE DIFFERENTIAL EXPANSION
Average free expansion or contraction in mm per metre of length

Material	10°C	20°C	30°C	40°C	50°C	60°C	70°C
Aluminium alloys average	0.236	0.472	0.708	0.944	1.180	1.416	1.652
Brass	0.188	0.375	0.563	0.750	0.938	1.125	1.313
Brickwork	0.056	0.112	0.168	0.224	0.280	0.336	0.392
Concrete	0.142	0.284	0.426	0.586	0.710	0.852	0.994
Copper							
Glass							
Lead	0.10	1.020	1.530	2.040	2.555	3.060	3.570
Stainless Steel	0.178	0.356	0.534	0.712	0.890	1.068	1.246
Structural Steel	0.121	0.242	0.363	0.484	0.605	0.726	0.847

Adapted from table prepared by Comalco Industries Limited.

Heat Transfer
Heat transfer is an important factor in relation to building materials. This phenomenon has several aspects which need to be understood.

The sun produces waves known as light, infra-red, ultra-violet and heat.

High-temperature heat, such as in the sun's rays and electrical radiators, can be radiated across spaces and absorbed or reflected by items on which the heat falls. This type of heat, as with solar heat, is of relatively short wavelength and can be largely reflected by shiny or white surfaces.

Low-temperature heat, such as produced by items warmed by the sun's rays (e.g. the atmosphere), re-radiate their heat at longer wavelengths, which are not so readily reflected as the short waves. Consequently, some materials, such as paint or light coloured materials which are effective against solar radiation, are less effective against atmospheric or other long wave heat.

Absorption, Emissivity, Reflectance
To quantify and tabulate thermal qualities therefore require different figures for solar radiation or terrestrial radiation. Relevant names used are Absorptance and Reflectance characteristics. Also in this context the Emmissivity of a heated material denotes the ability to re-radiate heat absorbed. For any given wavelength/Material Emissivity and Absorptance are equal.

The sum of Reflectance and Absorptance should always equal one.

An indication of how surface colour affects the performance of materials in this regard is given by the following tabulation.

TABLE 3
SURFACE REFLECTANCE/ABSORPTION

Surface	Solar Radiation Short Wavelengths		Terrestrial Radiation Long Wavelengths	
	Absorption Emissivity	Reflectance	Absorption Emissivity	Reflectance
Aluminium	.05	.95	.05	.95
Fresh Whitewash	.12	.88	.90	.10
White paint	.20	.80	.90	.10
Galvanised steel	.25	.75	.25	.75
Light coloured paint	.40	.60	.90	.10
Concrete	.60	.40	.90	.10
Dark coloured paint	.70	.30	.90	.10
Black paint	.85	.15	.90	.10

The implications of surface colour for performance criteria on surfaces performing solar insulating functions in hot climates should be obvious.

Thermal Conductivity (k) Refers to Unit thickness
The thermal conductivity of a material defines its rate of heat transfer. A material with a k value of 1 will transmit heat at the rate of 1 watt for every degree of temperature difference between opposite faces of a cube with 1 metre sides.

In descending magnitude figures for numerous materials, 'k' values are given below. These figures can vary depending on the density and moisture content of the material.

TABLE 4
GRADUATED CONDUCTIVITY COMPARISON

Granite	4.220	Caneite	0.060
Sandstone	1.150 to 2.300	Foamglass	0.054
Brick	1.150	Mineral Wool	0.038
Glass	1.050	Cork	0.038
Concrete	1.000 to 1.500	Fibreglass (GRP)	0.034
Still Water	0.667	Polystyrene	0.031
Blast Furnace Slag	0.250	Expanded Ebonite	0.028
Wood	0.144	Still Air 10oC	0.024
Strawboard	0.090	Polyurethane	0.021

Thermal Resistance

Heat can be readily transmitted through some materials (conductors), notably metals, and can be resisted by other materials, e.g. glass, fibre batts, polystyrene foam, timber (insulators). The measure of this 'quality' is known as 'Thermal Resistance' and is expressed as the energy required to raise a square metre of the material up one degree centigrade. Obviously, this will vary with the thickness of the material, so for comparative purposes it is useful to tabulate results on the basis of some commonly used dimensions for building materials.

Accurate figures for individual products should be available from the manufacturers or marketers, but for rough comparative purposes the following tabulation may be useful.

TABLE 5
THERMAL RESISTANCE OF BUILDING COMPONENTS

	W/mK	m2 °C		W/mK	m2 °C
25mm Aluminium	0.00012		150mm Glass fibre batt	3.38	
100mm Brick	0.14		25mm Polystyrene foam	0.76	
200mm Concrete	0.16		25mm Polyurethane foam	1.06	
Single Glass (window)	0.16		25mm Timber (pine)	0.14	
Double Glazing (window)	0.32		Wall well insulated	2.7 to 3.4	
Glass fibre batt (90 mm)	1.96				

This table shows how little heat is needed to heat the metal (25mm aluminium) compared with 25 mm polystyrene, polyurethane or timber.

The figure for 200mm concrete must be carefully compared alongside the equivalent figure for thin sheet window glass. This shows that glass itself is a good insulator, but because it is used in very thin sheets compared with masonry and concrete, glazed areas are usually comparatively poor insulators.

TABLE 6
THERMAL RESISTANCE

	Nominal Thickness L (mm)	Resistance R (m²K/W)
Masonry Wall Materials		
Bricks — Common	110	0.153
Face	110	0.085
Concrete — Bricks — dense	110	0.066
Blocks — cored dense	100	0.126
	200	0.176
Concrete dense, solid 2300 kg/m³	150	0.094
Lightweight 1800 kg/m³	150	0.212
1100	150	0.416
300	150	1.724
Stones — Granite	150	0.051
Sandstone	150	0.087
Marble — Limestone	150	0.1
Plaster — cement or gypsum)	13	0.018
+ sand aggregate)	19	0.026
Flooring Materials		
Carpet + fibrous underfelt		0.367
Ceramic tile	25	0.014
Cork tile	6	0.14
Linoleum	3	0.015
PVC tiles or sheet	2	0.003
Rubber tiles	3	0.014
Timber — Softwood	19	0.173
Hardwood	19	0.10
Roofing Materials		
Asbestos Cement — Deep corrugated	7	0.027
Built up roofing	9	0.059
Slate	13	0.009
Sheet metal — all types		Negligible
Strawboard (compressed)	50	0.575
Tiles — terra cotta		0.018
Wood shingles		0.166
Wood wool slabs and cement	50	0.61
Cladding Materials		
Fibre cement sheet	4.5	0.017
	9	0.035
Glass	3	0.004
	6.5	0.008
Hardboard — (Weathertex)	9.5	0.069
Metal Siding — foam backed	9.5 av.	0.238
Timber weatherboards — softwood	16 av.	0.145
hardwood	16 av.	0.084
Lining Board, Sheets, Panels		
Fibre Cement Products		
Standard sheets	4.5	0.017
	9	0.035
Autoclaved sheets	6	0.024
Fibrous Plaster	9	0.036
Standard sheet	9.5	0.026
Fyrated sheet	16	0.044
Gypsum Plasterboard	10	0.057
'Fyrcheck'	13	0.074
	16	0.091
Hardboard	3.2	0.022
	5.5	0.038
Particleboard	10	0.087
	25	0.217
Plywood	4	0.025
	9	0.055
	19	0.177
Softboard	9.5	0.19
	19	0.388
Softwood Lining Boards	12	0.109
Insulation Materials		
Fibreglass or Rockwool	50	1.2
Batts or Blanket	100	2.4
Acoustic ceiling panels	25	0.70
Corkboard		

Thermal Capacity

Heat absorbed by materials is normally stored until ambient temperatures drop below the temperature of the material. The stored heat is then re-radiated until a balance is achieved. This phenomenon can be important in the thermal comfort of buildings. Wisely chosen material can promote energy conservation and the comfort of occupants.

This ability to store heat is called thermal capacity and is measured as the amount of heat required to raise the temperature of a unit volume or unit weight by one degree.

Water has the highest thermal capacity of common materials. In general, thermal capacity is roughly proportional to mass - large quantities of dense materials can store large quantities of heat. Loosely arranged and porous materials store little heat (e.g. concrete and brick have high thermal capacity and timber low capacity).

The measurement of thermal capacity is the heat required to raise one unit by one degree. For units of mass, it is expressed as Kilojoules per kilogram per degree centigrade:

$$\frac{KJ}{m^3\ {}^oC}$$

For units of volume, it is Kilojoules per cubic metre per degree centigrade:

$$\frac{KJ}{M^3C}$$

The following table indicates how little heat is stored by porous materials such as timber and mineral wool compared with the denser materials and water, but is of little use in actual building situations.

TABLE 7
THERMAL CAPACITIES

	Per Unit Mass KJ Kg °C	Per Unit Volume KJ m3 °C
Water	4.19	4160
Steel	0.50	3960
Stone	0.88	2415
Concrete	0.88	2080
Brick	0.84	1680
Clay soil	0.84	1350
Wood	1.89	940
Mineral wool	0.84	27

For practical purposes, it is much more convenient to tabulate such information in terms of commonly used construction systems, such as walls, on a per square metre basis.

In assessing or calculating the performance of complete elements of construction, such as stud-framed walls, cavity brick walls, roof and ceiling construction, account must also be taken of built-in cavities containing air and the degree of ventilation available in such cavities. This can then be expressed as a Thermal Resistance - R (Heat Resistance Units) for the complete assembly.

Table 7 lists the ratings for components of building elements, such as walls, floors, etc., but the total R for complete elements is not simply the sum total of all components, and needs to be calculated by specialists.

A useful discussion paper on this topic is BDR1 Technical Note 8A - Design of Full Brick Houses. This contains some comparative assessments, graphs, diagrams and tabulations suited to some domestic situations.

Refer also to

NZS 1340 - 1970 - Thermal insulating materials for buildings. BS 3958 - Thermal insulating materials.

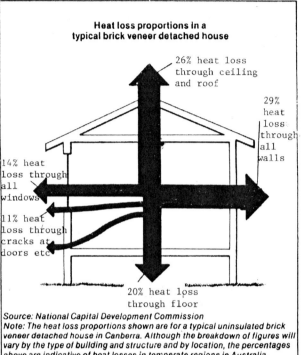

Heat loss proportions in a typical brick veneer detached house

26% heat loss through ceiling and roof

29% heat loss through all walls

14% heat loss through all windows

11% heat loss through cracks at doors etc

20% heat loss through floor

Source: National Capital Development Commission
Note: The heat loss proportions shown are for a typical uninsulated brick veneer detached house in Canberra. Although the breakdown of figures will vary by the type of building and structure and by location, the percentages above are indicative of heat losses in temperate regions in Australia.

Domestic Heat Loss Diagram

From the diagram above it is clear the detached house is not highly efficient in terms of heat loss or gain, having great areas exposed to the elements. The commonly built multi-storey apartment or two storey terrace type house with two external walls and minimal roof area produces a thermal performance which gives much less variable seasonal temperatures. The adjacent tenanted spaces also contribute to this stability.

References
See Introduction page 2 for Thermal References and Standards

CSIRO Publications
BTF 4 & NSB 163   Thermal Insulation
NSB 176A   Sarking & vapour barriers

112

TABLE 8
GENERAL THERMAL PROPERTIES OF SOME COMMON MATERIALS

Adapted from N. H. Hassall's St Regis-ACI Handbook et al.

Material	Density Kg/m3 d	Specific Heat s	Conductivity W/(rnK) k	Coefficient of thermal expansion Strain °C
Air (Dry 30°C)	1.2058	.241	.026	-
Aluminium	2560	.215	200	23-24
Asbestos Cement	1600	.20	.650	
Brass	8400			
Brickwork	1600	.20	1.150	
Corkboard	160	.43	.038	
Concrete aerated	320	.21	.086	12
Concrete dense	2320	.21	1.500	12
Copper	8700			
Fibreboard/ Caneite	270	.36	.060	
Glass	2480	.16	1.050	9
Granite	2640	.195	4.220	11
Hardboard	1120	-	.201	
Lead	11340		3.476	29.3
Limestone	2500			
Plasterboard	990	.25	.173	
Plywood				
Polystyrene (foamed)	16		.031	70
Polyurethane (foamed)	32		.021	
Sandstone	2000		1.150 to) 2.300)	12
Steel - structural	7850	.12	50	12
Tiles (clay)	1920	.22	.836	
Timber	650 to) 1100)			
Vermiculite (exfoliated loose)	80 to) 110)	.21	.065	-
Water	1000	1.00	.667	-
Wood (western red cedar)'	480	.45	.144	-

NOTE: The above figures are generalisations. For accurate calculations, details of alloys, ingredients and sample testing may be necessary on materials being used.

Fire Hazard Indices

Formal testing of materials and construction systems has led to the establishmnent of Early Fire Hazard Indices and related information. The tests give indices for ignitability, spread of flame, heat evolved and smoke. These are published for various groups of materials as Notes on the Science of Building, and the following are relevant in this aspect:

NSB
66 Fire aspects of combustible wallboards and finishes
98 Lightweight fire-resisting construction
136 Fire doors, fire shutters and fire windows
137 Fire hazards of furniture and furnishings
142 Fire hazards in the home

Toxic Gases

Escape from large buildings in the event of fire has become a major concern of fire-fighting authorities and building regulations. It is now realised that the most serious threat to entrapped humans is smoke and inhalation of toxic gases generated by many materials when subjected to heat.

Research work is being carried out to determine the degrees of danger involved. To date no generally accepted standards have been formulated, but a series of tests conducted by American laboratories indicate that traditional organic materials, such as timber, cotton and woollen products used inside buildings, produce highly toxic gases in fire.

Modern plastic materials have come under close scrutiny in this regard and these tests suggest that no generalisation is warranted as some polysters approximate cotton and wood in toxicity, while others like PVC, polyurethane foam and polystyrene are far less dangerous.

Further references worth noting are
NZS - 4502 - Glossary of terms associated with fire.
1. The phenomenon of fire.
2. Building materials and structures.
BS 4422 Glossary of terms associated with fire.
BCA - Specification A2.4 - Early Fire Hazard Test for Assemblies.

BDRI research graph comparing thermal performance of full brick, brick veneer and weatherboard construction.

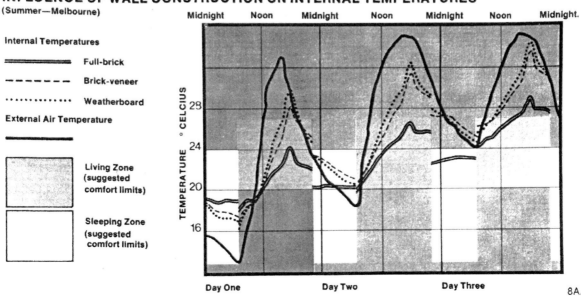

INFLUENCE OF WALL CONSTRUCTION ON INTERNAL TEMPERATURES
(Summer—Melbourne)

8A.

Comparative testing of some materials to AS 1530, Part 3 — Early Fire Hazard Test

Material	Ignitability Index (0–20)	Spread of Flame Index (0–10)	Heat Evolved Index (0–10)	Smoke Developed Index (0–10)
EBS	12	0	3	5
Australian Hardwood (bare)	14	6	7	3
Australian Softwood (bare)	16	9	7	3
Oregon	13	6	5	3
Bluegum	11	0	3	2

Source: EBS: AWTA Test Report No. 9-96156
Other materials: EBS Notes on the Science of Building NSB66.

Some comparative Early Fire Hazard Indices.

FIRE RESISTANCE RATING AND FIRE RESISTANCE LEVEL

The Building Code of Australia has introduced new terminology and a comprehensive tabulation of Fire Resistance Levels (FRL) in minutes, for commonly encountered materials and systems. These tables should be referred to for all buildings coming under the Code.

The following notes and diagrams on this topic serve as useful explanatory notes.

Fire-Resistance Ratings

Fire-resistance ratings are determined by laboratory fire testing of an assembled component so that no one material can have a fire rating on its own.

However, as . brickwork and reinforced concrete are so widely used to develop fire-resisting construction, they are generally accepted as having ratings according to the thickness of the wall. In some cases, the surface finish can improve the rating of a brick wall.

The following diagrams from BDRI pamphlet on Fire Ratings of Clay Brick Walls is self-explanatory and shows the rating in hours for each type of construction.

The following requirements must be observed to achieve the ratings given:

Net Volume

The bricks used must have a net volume of not less than 75% of their gross volume and any holes in the bricks must be in the plane of the wall. No other restrictions need to be placed on the shape, number or arrangement of holes since the recommended ratings, are conservatively based so as to apply to the most adverse combination of material properties and perforation pattern.

Maximum Slenderness Ratio

In order to achieve the fire ratings listed, the slenderness ratios of walls shall not exceed:

18 for cavity walls in which one leaf is loadbearing and one leaf is non-loadbearing;

20 for cavity walls in which both leaves are loadbearing; and for loadbearing single- and double-leaf walls; or

27 for non-loadbearing walls.

Calculation of Slenderness Ratio For Loadbearing Walls

The slenderness ratio of a loadbearing wall shall be the effective height divided by the effective thickness both calculated in accordance with the SAA Brickwork Code AS1640-1974, except that for the purposes of fire ratings only:

(a)   the actual height of the walls measured between supports shall be used when calculating the effective height;

(b)   the effective thickness of the walls shall not be deemed to be increased by piers, buttresses or intersecting walls;

(c)   the effective thickness of rendered walls shall be the actual thickness of the wall without render; and

(d)   the effective thickness of walls not uniform in thickness throughout their height shall be the thickness of the thinnest part.

For Non-Load bearing Walls

The slenderness ratio of non-loadbearing walls shall be calculated by dividing their actual height by their effective thickness determined in accordance with the above conditions.

Tying and Bonding

Cavity walls and solid walls with more than one leaf must be tied and bonded in accordance with AS1640-1974.

Thickness of Render

Renders must have a minimum thickness of 19 mm on each face of the wall.

Some regulations vary fire-resistance -ratings depending on whether the components are loadbearing or purely partition walls. References should always be checked for compliance with local conditions,

For reinforced concrete work the rating depends on:

a)   whether. the member will be exposed to fire on the underside (e.g. in beams or slabs) or on a vertical face as in columns and walls,

(b}   on the thickness of cover over the steel reinforcement,

(c)   whether the concrete is prestressed or post-tensioned.

These ratings are usually defined in Standard Codes which should be referred to.

AS 3600 - Code for Concrete Structures.
AS CA 3- Prestressed Concrete Code.

114

FIRE RESISTANCE RATING (hours)

1

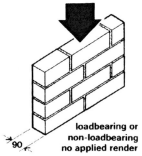

loadbearing or
non-loadbearing
no applied render

90

NOTES

Non-loadbearing walls only.

Loadbearing or non-loadbearing walls.

1·5

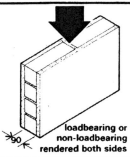

loadbearing or
non-loadbearing
rendered both sides

90

loadbearing or
non-loadbearing
no applied render

110

2

loadbearing or
non-loadbearing
rendered both sides

110

3

2 leaves non-
perforated bricks
on edge

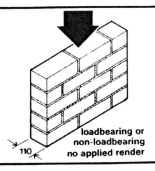

loadbearing or
non-loadbearing
no applied render

76 76
160

4

loadbearing or non-loadbearing
no applied render

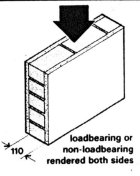

190

loadbearing or non-loadbearing
no applied render

230

loadbearing or non-loadbearing
rendered both sides

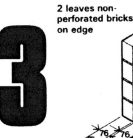

non-
perforated
bricks on edge

90 76
175

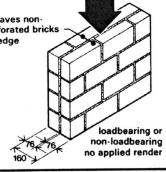

90 90

non-loadbearing only
no applied render

non-perforated
bricks on
edge

76 110

non-loadbearing only
no applied render

110
110 non loadbearing or
one or both leaves loadbearing
no applied render

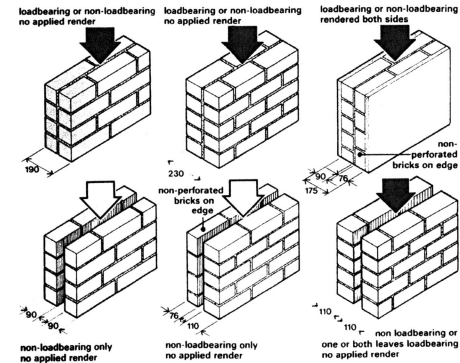

ACOUSTICAL PROPERTIES

The acoustical performance of a material is integrally related to the space in which it is used and to other adjacent materials. The relationship of sound to a material varies with the frequency of the sound (pitch) and whether the sound is airborne or caused by impact.

Analysis of acoustical problems is a very complex task, but architects and others often need some rough comparative figures for preliminary design purposes or for considering alternative materials related to a problem.

An important element in the acoustical property of any material is its density. Resistance to airborne sound transmission through the material increases with mass. Composite types of construction, however, incorporating special acoustical materials can produce good sound resistance without excess weight. Impact sound is best absorbed by soft materials which are usually comparatively lightweight.

A very useful reference containing numerous comparative tabulations and advice on acoustic isolation of spaces is the booklet `Noise Isolation with Lead Sheet' (1981) produced by Broken Hill Associated Smelters, 55 Collins Street, Melbourne 3000. This and other information regarding lead's acoustical applications is available from the Lead & Zinc Development Association, 95 Collins Street, Melbourne 3000.

The booklet contains instructions on how lead sheet should be used for acoustical purposes, as well as graphs comparing conventional wall, floor, ceiling and partition systems, and how lead sheet can improve the acoustical performance of each.

The Sound Control Guide published by Hardboards Australia Ltd. also contains numerous applications- and acoustical performance graphs for stud-framed construction.

For detailed information, books on acoustics should be consulted, but the following tables give a few other useful comparisons.

Sound Reduction Indices

Sound is measured by instruments which can break the incident sound up into the various component frequencies. The intensity of sound is measured in `decibels' (symbol dB). The performance of various subdividing components have been laboratory tested to determine their sound-reduction ratings. These are also expressed in dB units.

It must be remembered that acoustics parallel thermal performance in regard to the role of windows, doors or other changes in material in a barrier. A highly efficient wall barrier can be negated by an inefficient door opening or bypass route.

The following table gives a very brief indication of some indices. More detail is available from Parkin - Acoustics, Noise and Buildings.

TABLE 11
SOUND REDUCTION INDICES

Type of Partition	Average dB Reduction	Reduction	Average dB Reduction
110 mm Brick Wall plastered	45	Window closed, 4 mm glass	22
220 mm Brick Wall	50	Window, double glazed 200 mm air space,	
150 mm dense concrete wall	47	tightly sealed	39
Timber floor and) plasterboard ceiling)	34		
125 mm concrete floor	45		

TABLE 10
COMPARATIVE WEIGHTS OF BUILDING MATERIALS IN SITU

Material	Nominal Thickness mm	Nominal Mass Kg/Sq. metre
Aluminium sheet	1	2.7
Asbestos cement	6	9.7
Brickwork	110	200
Chipboard	10	6
Concrete	100	230
Concrete blockwork	200	Solid 400
		Hollow 200
Glass	6	14.6
Hardboard	3	3.4
Plasterboard	10	8.3
Plaster, gypsum or lime	10	16.3
Plywood	3	1.47
Slate	25	72
Steel sheet	1	8
Stone (average)	25	52
Strawboard compressed (Stramit)	50	18.5
Timber -- seasoned		
Hardwoods average	20	20
Softwoods average	20	10

References

The Building Code of Australia adopts by reference Standard Code AS1191: 1985 method for laboratory measurement of airborne sound transmission loss of building partitions: and AS2107: 1977 Code for ambient sound levels within buildings.

Graph indicating the improved sound insulation possible in stud framed walls by using lead sheet. Reproduced from BHAS - Noise Isolation with Lead Sheet.

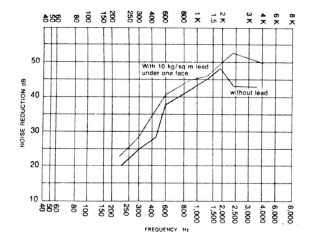

TIMBER STUD WALL

100 mm hardwood timber studs at 457 mm centres
(a) Clad both sides with 13 mm plasterboard (BHAS Test 1963).

Thickness	126 mm
Surface Mass	20 kg/sq m
Noise Reduction	37 dB

(b) Clad one side with 13 mm plasterboard and the other side with 10 kg/sq m lead sheet adhesively bonded to 13 mm plasterboard (BHAS Test 1963)

Thickness	126 mm
Surface Mass	30 kg/sq m
Noise Reduction	43 dB

LaVergne, TN USA
10 March 2011
219623LV00001B/71/P